How Not to Grow Old

By the Anti-Aging Research Group

AF882436

Editor: Amber Lin

Authors:
Alex Chen, Amber Lin, Angela Lin,
Ashley Zhang, Jacqueline Ni, Jonathan Chen,
Kimberly Liang, Lauren Cheng, Sunny Peng

PREFACE

In recent years, old age has become an increasing problem due to its influence on age-related diseases. According to the World Health Organization's International Classification of Diseases, aging is "caused by biological processes which persistently lead to the loss of organism's adaptation and progress in older ages", a change from the previous consensus of aging being a natural phenomenon and being unable to be treated.

Over the past few decades, more people have started paying attention to slowing down aging, creating a market for anti-aging supplements derived from natural compounds, such as Nicotinamide Mononucleotide (NMN). In addition, there is new research in the field of age-related diseases such as cardiovascular disease, diabetes, cancer, and Alzheimer's disease, and finding organic ways to alleviate these diseases, such as a vegan diet. From current research, we can devise better treatment options for patients that impact the origin of the disease, aging, rather than a one-time solution for the worsening of an age-related disease.

How Not to Grow Old is written by us, the Anti-Aging Research Group. We are an organization of high schoolers, and our goal is to teach people of all ages on how to have a healthy lifestyle. Our book focuses on one topic per chapter: NMN and other supplements, veganism and cardiovascular disease, veganism and diabetes, veganism and cancer, and Alzheimer's disease, all in relation to aging. In each chapter, the topic will be explained and solutions to each topic will be expanded on. In addition, research conducted by the authors of this book will also be touched on in certain topics. Our book highlights the importance of involving youth in scientific research of the anti-aging field and beyond.

I hope that through this book, people of all ages will understand aging better and are able to improve their health and quality of lives. Happy reading!

– Amber Lin, president of the Anti-Aging Research Group

TABLE OF CONTENTS

CHAPTER 1

NMN AND OTHER SUPPLEMENTS

Part 1: What are Anti-Aging Supplements?

Anti-aging is a novel field of medicine, will continue to grow, and studies have already shown that molecules could be used without the intake of artificial supplements. Anti-aging supplements, which have been available since the 1980s when Dior introduced liposomes in their lotions, have garnered increased scientific attention for their potential impact. Beyond merely aiming for a more youthful appearance, these supplements can be used to extend a healthy lifespan. [1] Among the well-known mechanisms, they involve mitigating the effects of harmful molecules called free radicals. These unstable molecules, known as free radicals, can wreak havoc on our bodies, which can potentially lead to serious conditions such as heart disease and cancer. Thankfully, antioxidants, like vitamins C, E, and beta-carotene, which can help neutralize these harmful effects, can be easily incorporated into our diets.

Vitamin C, found in citrus fruits, strawberries, bell peppers, and broccoli, is mostly recognized for its ability to counteract free radicals. These highly reactive molecules are known to damage cells, and can accelerate premature aging. Vitamin C's effects also extend to things such as reducing wrinkles by promoting collagen synthesis, a protein crucial for skin firmness. In other words, Vitamin C can drastically improve skin elasticity to achieve a more vibrant complexion. [3] Additionally, there is research that indicates that Vitamin C effectively decreases wrinkle levels over twelve weeks! The natural decline in collagen production with age points out the importance of Vitamin C in boosting synthesis and protecting skin proteins, which is vital for maintaining skin health and combating aging signs. [4] Moreover, Vitamin C is a crucial dietary addition that is responsible for increasing skin vibrancy and elasticity along with increased health and ultimately is a powerful natural supplement for anti-aging.

In addition to Vitamin C, another essential anti-aging nutrient is Vitamin E, which keeps the skin healthy by combating oxidative stress while also shielding against harmful UV radiation. Additionally, it has anti-inflammatory properties, with many reports detailing treatments where Vitamin E (along with other vitamins) successfully helped treat skin conditions. [5] Vitamin E is often believed to possess anti-inflammatory properties, meaning that it can help reduce redness throughout the body. Even more surprisingly, Vitamin E's antioxidant nature, which is enabled by its fat-soluble properties, allows for storage within the body. [6] Furthermore, because of its anti-inflammatory properties, Vitamin E is often seen as a holistic approach to treating skin conditions and benefits skin health.

However, antioxidants are just one facet of anti-aging supplements. Other examples encompass Quercetin, Curcumin, Omega-3 Fatty Acids, Green Tea Extract, Metformin, and Probiotics. Quercetin is a supplement that promotes healthy aging & potential for anti - inflammatory responses and can be found in fruits, vegetables, black tea, nuts, and wine. [7] Quercetin is known for its anti-inflammatory effects, inhibiting the production of molecules such as cytokines and lipoxygenase, ultimately reducing inflammation. Curcumin, another supplement, also has anti-inflammatory and antioxidant properties and is found in turmeric. [7] Similar to quercetin, curcumin is beneficial in reducing inflammation and oxidative stress.

Beyond antioxidant and anti-inflammatory advantages, these supplements contribute to heart and brain health, support youthful skin, fortify the immune system, facilitate nutrient absorption, etc. [7] Overall, chaining your diet to incorporate more anti-oxidants will benefit your health and cause less inflammation to occur.

Fortunately, the increasing understanding of anti-aging supplements is due to continuous efforts and advancements of the scientific community in this field. There are numerous researchers dedicated to advance the field, with some who have even "pushed" the boundaries of knowledge in this field, including Shinya Yamanaka, a prominent Japanese scientist renowned for his work with induced pluripotent stem cells. In 2006, Yamanaka achieved a breakthrough by successfully reprogramming adult mouse cells into pluripotent stem cells. The reprogramming was done through the introduction of four key transcription factors, marking a shift in stem cell research, and providing a promising alternative to the ethically complex use of embryonic stem cells. His discovery proved crucial in widespread anticipation for the potential of regenerative medicine. Yet, while Yamanaka's accomplishment was indeed a landmark discovery, it represents only a crucial initial step in a long journey toward comprehending the intricate mechanisms of aging and realizing effective anti-aging treatments. This field needs comprehensive research, rigorous experimentation, and ongoing collaboration across disciplines to pave the way for genuine breakthroughs. [8-9]

Shinya Yamanaka's pioneering work in induced pluripotent stem (iPS) cells commenced with a groundbreaking achievement involving adult mice. By successfully inserting four specific genes into the nuclei of fibroblast cells—cells integral to connective tissue—the adult mice cells were able to transition into pluripotent ones. A year later, the scientist replicated his success with human cells with the use of a retrovirus. The retrovirus carried a *c-Myc* gene, which led to a discovery. This gene was identified as a key factor responsible for the transformation of adult human cells into their pluripotent equivalents. However, the use of the retrovirus led to a major consequence: the development of tumors when the modified cells were transplanted into mouse subjects. With this limitation in mind, the scientist created adult human noncancerous iPS cells without the use of the retrovirus all together and leading to another breakthrough in 2008. [10]

In addition to Shinya Yamanaka's work in pluripotent stem cells, Dr. Thomas Rando's work focused on aging along with creating a research program centered on muscular dystrophies, tissue repair and stem biology. [11] In his works, Dr. Rando has shown that although quiescence was labeled as an unimportant stage in cells; however, recently, there has been some research that demonstrates that quiescence is an important part for longevity. Additionally, his stem laboratory discovered groundbreaking studies discovering that old tissue could rejuvenate if exposed to young blood. His work was fundamental for novel treatments for neurological conditions such as Alzheimer's disease. [12]

The works of Shinya Yamanaka and Dr. Thomas Rando, along with many other scientists, have paved the way in understanding and discovering the effects of aging and how we could reverse the potentially dangerous process. Through the scientists' research mentioned above, it is demonstrated that anti-aging properties can be shown because of how different molecules can react in the body to help lengthen the telomeres in chromosomes. Additionally, adopting a diet rich in antioxidants demonstrates that natural chemicals such as Vitamin C and other antioxidants can help with the anti-aging process.

Part 2: Resveratrol

On his 60th birthday, Donn Young was diagnosed with prostate cancer [13]. After getting a prostate-specific antigen test, which measures the prostate-specific antigen levels in the blood, the levels were at 78 [13], while healthy levels are at 4. Prostate cancer, which is the fourth most commonly diagnosed cancer worldwide, is a currently incurable disease and is often diagnosed in the senior population [14], as in individuals older than Donn Young. The diagnosis of prostate cancer, among many others, shows how aging is a disease [15]. Currently, research has been done to find solutions, especially using natural compounds as treatment options to prevent aging, such as resveratrol, a supplement that has been researched to have anticancer effects as well as anti-aging effects.

Resveratrol is a natural polyphenol and antioxidant that was first isolated from *Veratrum grandiflorum*, the root of white hellebore, by Taaoka in 1939 [16]. Unlike many other drugs, resveratrol reaches many different molecular targets, which allow it to access many cancer pathways at the same time [16]. Since resveratrol is a phytochemical, a natural compound in plants, it modifies signaling pathways without toxicity [17]. In addition, resveratrol has been researched to reverse the drug resistance in cancer cells that may develop after being exposed to certain chemotherapeutic agents [17]. One way resveratrol inhibits cancer is through the regulation of genes that affect cancer cell proliferation or tumor cell growth. For example, an extract of a Cassia plant, whose active ingredient is resveratrol, inhibits cyclooxygenase (COX-1), causing a reduction in tumor cell growth and a reduced risk of colorectal cancer [4]. In addition, resveratrol changes the balance of cyclins and CDKs, which are cell cycle regulators, causing inhibition of the cell cycle in cancer cells [17]. Resveratrol increases the levels of cyclin A and E, as well as down-regulating the expression of cyclin D1, D2, and E, which led to a reduction of skin cancer A431 cells [17]. Furthermore, resveratrol has been researched to inhibit TCDD-induced expression of cytochrome P450 1A1 and 1B1, which can cause catalytic actions in human breast cells and lead to cancer [17]. Therefore, resveratrol limits pro-carcinogen activity in order to prevent cancer. Also, resveratrol arrests cell cycles by activating the p-53-dependent signaling pathway, which causes tumor suppression [17]. Moreover, resveratrol induces apoptosis by inhibiting the PI3K/Akt/mTOR pathway [17]. Equally importantly, resveratrol inhibits signal transducers of transcription 3 (STAT3). STAT3 promotes the proliferation of tumor cells [17]. Finally, through the regulation of MMPs, TIMPs, EGFRs, and VEGFs, resveratrol causes anti-tumor progression activity [17]. In order to cause tumor progression, genes are first mutated in cells. Then, cancer cells undergo metastasis by destroying the extracellular matrix using matrix metalloproteinases (MMPs) [17]. Resveratrol downregulates MMP2 and MMP9, two types of MMPs. Also, tissue inhibitor metalloproteinase proteins (TIMPs) are natural MMP inhibitors [17]. Resveratrol has been researched to upregulate TIMP-1 protein expression in HepG2 cells, which are human hepatoma cells [17]. Because tumors need to grow new blood vessels through angiogenesis, they are stimulated by growth factors [17]. Resveratrol has been researched to inhibit VEGF, one type of growth factor [17]. In addition, resveratrol has been researched to inhibit EGFR and FGF2, which are markers for angiogenesis [17]. Additionally, resveratrol is able to activate NAD deacetylase sirtuin 1 (SIRT1), which promotes longevity and imitates caloric restriction, through activating AMP-activated protein kinase (AMPK) [16]. Through this method, resveratrol is also able to exert anti-aging effects on cells.

Some plants create resveratrol when they are wounded or attacked by fungi [18]. Proven by past studies, resveratrol has been shown to diminish cell proliferation and prevent strictures from forming in human patients with Crohn's disease [19]. It was utilized to increase apoptosis in fibroblasts and downregulate mRNA expression of procollagen type I and II in fibroblasts, which could be applied to patients with hypertrophic scar disorder [20]. Furthermore, resveratrol is able to target multiple types of cancer cells, including breast, skin, colon, and ovary cancer cells [21]. Resveratrol downregulates expressions of cyclin D2 and cyclin E in skin cancer cells to inhibit their development and upregulates p21 expression, a crucial gene for inhibiting cell growth [22]. Furthermore, resveratrol induces apoptosis in many pathways, such as the PI3K/Akt/mTOR pathway, and inhibits phosphorylation of P13K/Akt in multiple cancer cells [17].

Besides targeting genes in signaling pathways, resveratrol is able to inhibit oxidative stress, a main part of carcinogenesis, the start of forming cancers [17]. Reactive oxygen species (ROS) molecules react with DNA in addition to chromatin proteins, causing DNA damage [17]. Resveratrol finds radicals generated by cells and protects cells against lipid peroxidation, which would result in DNA damage by ROS and lead to cancer formation [17]. Therefore, resveratrol could be a possible chemoprotective agent.

Further, resveratrol is able to have anti-inflammatory effects, which can prevent cancer. In the specific case of Crohn's disease (CD), a disease of inflammation in the bowel area, fibrosis happens, and there is thickening of connective tissue [19]. This may cause devastating effects, including the development of intestinal strictures and fistulae that cause significant patient morbidity [19]. In a study on the *in vivo* effects of resveratrol on rats, resveratrol was able to decrease the number of rat intestinal smooth muscle cells (RISM) [19]. This was done by inducing apoptosis and cell cycle arrest, as well as decreasing collagen synthesis by the cells [19]. In addition, resveratrol decreases vascular smooth muscle cell hyperplasia in models of atherosclerosis through cell cycle arrest and apoptosis [17]. In other reports, resveratrol decreased collagen type I and III protein or mRNA expression in rat models of organ fibrosis [19]. In addition, resveratrol reduced inflammation and tissue damage in rat models of inflammatory bowel disease, and there were chemopreventive effects of resveratrol in models of colon, esophageal, and breast cancer [19].

However, resveratrol has its drawbacks in terms of pharmacokinetic abilities. 77–80% of resveratrol ingested has been found to be absorbed by the intestines [23]. Absorption is done through active transport as well as being bound to proteins in the blood. Because resveratrol is nonpolar, it has decreased bioavailability. In addition, it has low water solubility, causing resveratrol to be quickly excreted by the liver before it can be absorbed by the bloodstream. This makes resveratrol an inefficient drug because larger doses of resveratrol must be taken to achieve the same therapeutic level as water-soluble drugs.

Research has been conducted to improve its bioavailability and solubility. Research has shown that flavonoids such as quercetin increase resveratrol bioavailability, showing that changes to the diet could cause therapeutic blood and tissue levels of resveratrol to be achieved [23]. Resveratrol has been combined with cyclodextrins, the hydrophobic polysaccharide of starch, and improved its bioavailability by 60,000 times [23]. Another way was by nanoizing resveratrol to a particle size of less than 5 micrometers, increasing the surface area for absorption [23]. Nanoized resveratrol was shown to have higher concentrations in tissues like the liver and brain [23]. In terms of user consumption, resveratrol bioavailability was increased when taken with piperine [23], which is a compound found in pepper. It was researched to increase the resveratrol levels a thousandfold in rats [23]. In addition, when taken with other polyphenols found in red wine or plant polyphenols, resveratrol's bioavailability increased [24]. Furthermore, resveratrol's bioavailability increases when taken with compounds of the same enzymes [24]. Lastly, taking resveratrol sublingually has better bioavailability than taking it orally [24]. For example, grape seed soil could be used to encapsulate and deliver resveratrol.

Clinical trials conducted on resveratrol have shown that resveratrol has a range of anti-aging effects, such as improving memory in older adults. In a clinical trial, 24 healthy adults ranging from 50 to 80 years old were given 200 mg of resveratrol over 26 weeks with a 3-month resting period, compared to a control group of 23 subjects who took placebo pills [25]. Human subjects were tested before and after the experimental period on memory retention, blood parameters, and MRI imaging [25]. For the results, there was a significant increase in word retention after 30 minutes for the experimental group compared to the placebo group [25]. In addition, resting-state functional connectivity (FC) patterns [26], which are brain interactions that occur during an absence of tasks, were increased in the experimental group. In previous studies, there have been connections between memory increases and resting-state FC patterns [25]. Furthermore, results showed that resveratrol reduced glycated hemoglobin (HbA1c) levels in the experimental group [25]. Hemoglobin becomes glycated when glucose attaches to it, so HbA1C tests measure the sugar levels in the bloodstream over the past three months [27]. A glucose increase has been linked to lower cerebral blood flow and glycation of brain proteins, leading to worse memory [25]. Since the results show that resveratrol could improve and maintain memory retention in older adults [25], resveratrol has been shown to be an effective anti-aging supplement.

Besides memory, resveratrol has been shown to impact cardioprotective molecular targets that are involved with diseases in aging [27]. Resveratrol's anti-inflammatory properties have been a main component of this effect [27]. A clinical trial on resveratrol and cardiovascular disease showed that resveratrol improved systolic and diastolic function and reduced low-density lipoprotein cholesterol levels [27]. The deposition of fat is correlated with the blockage of vessels, leading to atherosclerosis [27]. Therefore, treatment with resveratrol could reduce the development of atherosclerosis. In addition, resveratrol reduces blood pressure, leading to a reduction in hypertension and heart disease [27]. Furthermore, in a clinical trial conducted on adult smokers, resveratrol was shown to reduce inflammation of the airways, leading to increased cardiovascular health [27].

Furthermore, resveratrol has been found to have an impact on diabetes. HbA1c levels, as mentioned earlier, can be a good indicator of whether patients have type 2 diabetes mellitus [28]. In a clinical trial of 62 patients, aged 30 to 70 years old, with type 2 diabetes mellitus, the experimental group took 250 mg of resveratrol daily along with hypoglycemic agents for three months, compared to a control group who took hypoglycemic agents only [28]. The experimental group patients had reduced HbA1c levels as well as lower blood pressure, cholesterol, and protein [28]. As diabetes risk has been linked to increasing age [28], resveratrol's role in regulating glucose levels makes it an effective anti-aging supplement. In another clinical trial on 10 overweight adults older than 65, human subjects were given either 1, 1.5, or 2 g of resveratrol daily over four weeks [29]. Human subjects were measured for their glucose and insulin under the curve (AUC), insulin sensitivity, insulin secretion, and endothelial function [29]. After the 4-week testing period, insulin sensitivity and post-meal insulin were decreased [29], indicating that resveratrol could be a deterrent to age-induced diabetes.

In conclusion, past research has shown the many anti-aging and anti-cancer effects of resveratrol, ranging from improving memory and reducing insulin sensitivity for type 2 diabetes mellitus to reducing oxidative stress. Topics of current research include improving the low bioavailability and low water solubility of resveratrol to increase resveratrol's efficacy. Due to resveratrol's beneficial anti-aging properties, more research needs to be done regarding clinical trials on resveratrol in order to determine its complete effects on anti-aging.

Part 3: The Secret to Eternal Youth:
The Anti-Aging Effects of NMN – Author Research

In the US, approximately 2.5 million people aged 65 or older simply die from old age every year. In addition, many life-threatening diseases, such as diabetes and heart disease, are age-correlated. For example, according to the Centers for Disease Control and Prevention, cancer and diabetes are the leading causes of death worldwide [30]. During childhood, grandparents died from age-related diseases. Saddened by experience and wanting to help more people live healthier, longer lives, the authors wanted to address biological aging to decrease the chances of getting age-related diseases. Scientists such as David Sinclair, a biochemist and professor at Harvard Medical School, have argued that old age is a disease and that it is not inevitable [31]. His research on rats has shown how, for example, NMN increases the NVC (neurovascular coupling) response in old mice, which is associated with the production of mitochondria, the organelle involved in cellular respiration that requires a continuous supply of reduced NAD+ to function. According to David Sinclair, taking supplements such as Nicotinamide Mononucleotide (NMN) could decrease biological age. NMN is a precursor of NAD+ (Nicotinamide Adenine Dinucleotide), a coenzyme that is essential for the functioning of many important enzymes, including those involved in cellular metabolism. NAD+ levels in the body decline with aging. Using David Sinclair's research as inspiration, the authors conducted a study on how NMN affects the feelings of youthfulness of human subjects over the age of 40.

In a research project conducted by the authors, a clinical trial was conducted on human subjects to measure the effects of NMN on feelings of youthfulness. A questionnaire with 10 questions, each on a scale of 1–5, was created. The questions focused on parameters of youthfulness that have been researched, such as energy level, happiness, well-being, sickness, pain, memory, mental alertness, focus, and curiosity. Then, human subjects that were over the age 40 were gathered, with the aim of half being male and half being female. There were twelve human subjects in the experimental group. Each experimental human subject was sent one bottle of liposomal NMN, purchased from the company Renue by Science. They will each take 250 mg (1 capsule) of NMN every day for 3 months. Then, they answered the questionnaire every two weeks in order to track the effects of NMN. Four human subjects were part of the control group and consumed gelatin capsules with protein powder. The control group was not aware that they were taking placebo pills since they were sent one bottle of gelatin capsules with the same instructions as the experimental subjects.

After three months, the data from the questionnaires were analyzed. Using the gathered data, the values of each human subject's questionnaire for every week were averaged to create Feelings vs. Time graphs for different groups of human subjects. In the Feelings vs. Time graph for male human subjects, there is an upward linear trend, with a 0.5429 increase each week. There is a t-score of around 3.7, and the p-value is around 0.007, which is statistically significant. In comparison, for male control human subjects, there is an upward linear trend with a 0.0071 increase each week. There is a t-score of around 0.11, and the p-value is 0.46 (not statistically significant). In the Feelings vs. Time graph for female human subjects, there is an upward linear trend with a 0.3548 increase each week. There is a t-score of around 1.35. In the Feelings vs. Time graph for female control human subjects, there is a downward linear trend.

There is a 0.2429 decrease each week, with a t-score of 0.11.

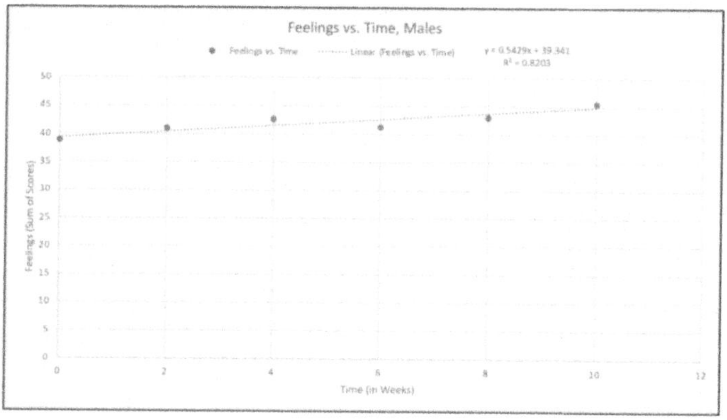

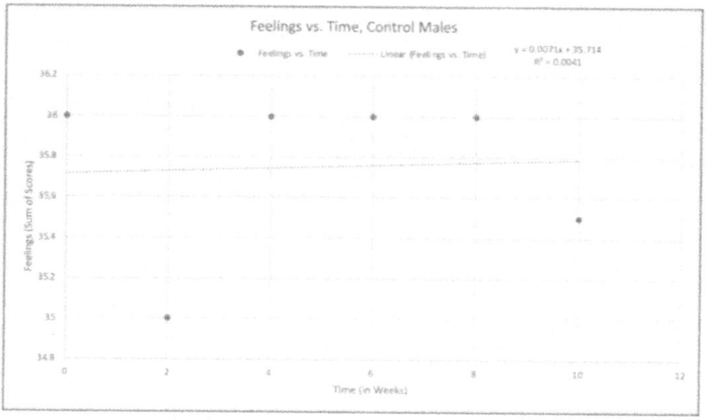

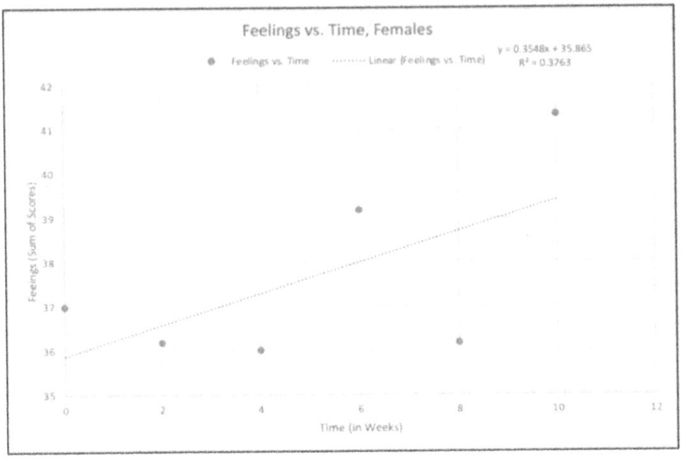

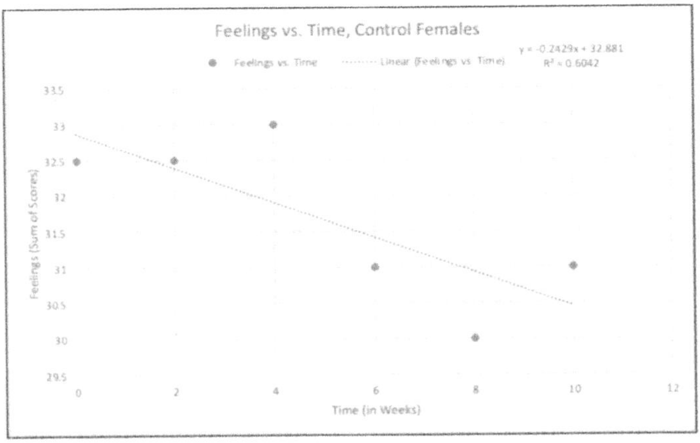

From research, it is known that NAD+ (Nicotinamide Adenine Dinucleotide) is a coenzyme present in all cells, and NAD+ levels decrease as we age. NMN (Nicotinamide Mononucleotide) is a precursor of NAD+, so when taking NMN, NAD+ is able to be produced, causing its levels to increase [32]. In addition, there is a marked increase in their feelings of youthfulness compared to control male and female human subjects, respectively. Moreover, the age averages of male and female human subjects are both much greater than their control counterparts. This could have caused the feelings of youthfulness to be lower.

Interestingly, for the control female human subjects, the results show that without NMN, their feelings of youthfulness will decrease. In a study on mice with diet-induced diabetes, female mice benefited more from their metabolic effects compared to male mice [33]. Calorie-reducing diets can cause NAD+ levels to rise and slow down aging, which could cause feelings of youthfulness to increase and produce the mentioned results [34].

In conclusion, the author hypothesis that NMN is able to increase the measurements of youthfulness is supported. There is a greater increase in average feelings of youthfulness weekly for male subjects compared to control male subjects, and a statistically significant value of <0.05 for male subjects. There is an upward trend for female subjects, while the control female subjects have a downward trend. In addition, there is a greater t-score for female human subjects compared to control female subjects. All four graphs show that there is a significant increase in the feelings of youthfulness in female and male subjects that took NMN continuously, compared to human subjects that did not take NMN.

These results will appeal to patients who want effective anti-aging results or be used in the medical industry to be applied to other anti-aging supplements. In addition, NMN can be used to treat patients with prediabetic symptoms since it increases muscle insulin sensitivity and insulin signaling in women [33]. NMN can be used to treat Alzheimer's and Parkinson's disease by creating more NAD+ molecules and fixing mitochondrial dysfunction, a main component of neurodegenerative disorders [35]. Because Alzheimer's and Parkinson's are age-related diseases, NMN is able to have anti-aging effects due to its impact on these diseases.

Part 4: Impact of NMN

As NMN, or Nicotinamide Mononucleotide, emerged as a substance with great potential, the numerous clinical trials demonstrated promising results. These initially began with uncertainty regarding whether NMN could reciprocate its effects from animal studies, such as mice, and instead, in humans. Previous research suggested that NMN promotes NAD+ production and has the possibility of preventing or improving many age-related issues. With a focus on this, these trials are targeted to expand on the existing findings about the effect of NMN on the human body.

In one double-blinded clinical trial conducted on 32 obese adults over the age of 55, participants were tested with MIB-626, which is a type of NMN, over the course of 14 days [36]. The participants were randomly assigned to three groups in which they received either 1000 mg of MIB-626 once or twice a day or were placed on placebo [36]. The researchers analyzed the NMN, NAD (nicotinamide adenine dinucleotide) levels, and NAD metabolome in each of the subjects after 14 days.

The results indicated that participants taking the NMN variant had significantly higher blood NMN concentrations compared to the placebo group [36]. The mean AUClast (area under the concentration-time curve) showed a 1.7-fold and 3.7-fold increase above the baseline in the once-daily 1000 mg and twice-daily groups, respectively. MIB-626 intake led to an increase in blood NAD levels among the participants, indicating that the oral intake of NMN contributes to higher concentrations of NAD in the blood [36]. Additionally, the blood levels of NAD metabolites were higher in the NMN-treated groups compared to the placebo groups on days 8 and 14, suggesting that NMN also contributes to increased levels of NAD metabolites [36]. The results did not demonstrate a correlation with sex, body mass index (BMI), or even age. Finally, it was found that very little NMN was discharged through urine, meaning that most of the NMN was likely used up within the body [36]. This clinical trial exhibits the relationship between NMN and NAD production, as well as the production of NAD metabolites. It also indicates the safety of NMN intake.

In another clinical trial, a similar trend of rising blood NAD levels was recorded. NMN supplements and placebo were given to 80 middle-aged, healthy adults. Either placebo, 300 mg, 600 mg, or 900 mg of NMN were distributed to the four groups of participants, divided randomly over the course of 60 days [37]. This trial aimed to evaluate blood NAD concentration as well as the safety and human body toleration of NMN supplementation. Physical performance (measured through a walking test), blood biological age, assessment for insulin resistance, and other subjective general health assessments completed by the participants were evaluated after the trial [37].

Following the trial, the primary objective of measuring blood NAD concentrations showed significant increases in all NMN-treated groups at day 30 and day 60 compared to both placebo and baseline ($p \leq 0.001$) [37]. The groups with higher NMN intake, specifically the 600 mg and 900 mg NMN groups, resulted in the highest NAD concentrations consequently [37]. Furthermore, the trial proved that NMN supplementation was well tolerated by all participants because no issues were reported on side effects or laboratory tests [37]. The physical performance task, a six-minute walking test, reflected that the groups receiving NMN had a significant statistical increase in their walking distances when compared to the placebo group at days 30 and 60 ($p < 0.01$) [37]. Among the NMN groups, the 600 mg and 900 mg NMN groups recorded the longest walking distances, suggesting improved physical performance of the participants. Moreover, while the blood biological age increased significantly in the placebo group, it remained unchanged in all NMN-treated groups, as it was previously hypothesized that there was a potential relation between NMN and biological aging. The insulin resistance assessment between the NMN-treated groups and placebo at day 60 did not show much of a difference; however, the subjective general health assessment scored higher in all NMN-treated groups [37]. Overall, this trial demonstrated the increase in blood NAD associated with NMN, improved physical performance, and had positive effects on subjective general health assessment. The trial also confirmed the safety and tolerability of NMN taken at a daily dosage of 900 mg [37].

Subsequent to this research, another clinical trial is being held to increase data on NMN supplementation in humans. This is due to the fact that numerous animal studies have demonstrated the preventive effects of nicotinamide mononucleotide (NMN) supplementation taken orally as being effective in alleviating age-related diseases. The study was conducted over a 12-week, randomized, double-blind, and placebo-controlled period with 36 healthy middle-aged participants [38]. Each of them received either 125 mg of NMN or placebo capsules twice a day. Post-trial, the levels of nicotinamide in the serum, a metabolite of NAD+, were significantly higher in the NMN groups than the placebo group [38]. Additionally, the pulse wave velocity values that indicate arterial stiffness tend to decrease in the NMN intake group [38]. Arterial stiffness refers to the stiffness or rigidity of the arterial wall. It gained increasing interest as it became a potential factor in the development of cardiovascular disease in adults [38]. The study concluded that NMN is effective in safely increasing NAD+ metabolism in adults and has potential for alleviating arterial stiffness.

In addition to the prior study that revealed the possible potential of NMN, a different study was held on accessing the aerobic influences of NMN on middle-aged adults. A six-week, randomized, double-blind, placebo-controlled, four-arm clinical trial was held with 48 young and middle-aged recreational runners of the Guangzhou Pearl River running team [39]. These participants are randomized into groups of four, categorized into a low-NMN-dosage group (300 mg/day), a medium-NMN-dosage group (600 mg/day), a high-NMN-dosage group (1200 mg/day), as well as the control group (placebo) [39]. Each group consists of 10 male participants and 2 female participants. Along with the NMN intake, each runner underwent 40–60 minute training sessions 5–6 times each week [39]. Cardiopulmonary exercise testing was conducted prior to and after the intervention to fully assess aerobic capacity [39].

The results of the study showed that the NMN groups had a much greater increase in oxygen intake, maximum oxygen uptake, power at the first ventilatory threshold, and power at the second ventilatory threshold [39]. However, there were no significant differences in the maximum oxygen uptake, oxygen pulse, VO2 related to work rate, or peak power between the groups after the six-week intervention [39]. Through the study, it was concluded that NMN supplementation can enhance aerobic capacity during exercise training in humans, improving oxygen use in skeletal muscles. The findings suggest a potential benefit of combining NMN supplementation with exercise training for improving cardiovascular fitness [39]. It concluded that NMN can benefit aerobic capacity in exercises such as running by improving oxygen utilization in skeletal muscles. [39] These findings suggest that we could potentially utilize NMN to improve the fitness of the cardiovascular systems.

NMN has also demonstrated more of its potential in a 10-week, randomized, placebo-controlled, double-blind clinical trial. This included postmenopausal women who were prediabetic and were either overweight or obese [40]. This is categorized by a body mass index exceeding 25.3 and falling below 39.1 kg/m2. A total of twenty-five women completed this trial; 12 were in the placebo group, and the remaining 13 belonged to the NMN group [40]. The first tests included dual-energy x-ray absorptiometry and magnetic resonance imaging; the hyperinsulinemic-euglycemic clamp procedure; and stable isotope glucose and palmitate tracer infusions to assess insulin sensitivity [40]. After taking NMN, metabolites produced by NMN increased in the following 10 weeks [40]. This suggested NMN treatment increased muscle NAD+ turnover. Additionally, following the 10-week treatment with NMN, the muscle insulin sensitivity was greater but remained unchanged in the placebo group. It is proven that NMN was specific to insulin sensitivity and did not affect other aspects of insulin resistance, demonstrating the specific benefits of the supplement [40]. NMN also proves to be improving on skeletal muscle biology. It increased skeletal muscle remodeling and regeneration by increasing insulin-stimulated mTOR phosphorylation and gene expression [40].

NMN and Future Applications

In the future, NMN could be developed and tested for many other illnesses and diseases that come with aging. One of which would be age- and diet-induced diabetes. NMN posed as a possible solution as it was tested on mice in this category, increasing the mice's insulin secretion, insulin sensitivity, and lipid profile in mice with type 2 diabetes [42]. At the same time, it improved hepatic insulin sensitivity as well as glucose tolerance in fat-induced diabetic mice. Another illness that NMN could possibly be a solution to is age-associated vascular dysfunction and oxidative stress [40]. NMN restored carotid artery endothelium-dependent dilation and nitric-oxide-mediated carotid artery endothelium-dependent dilation [40]. It also relieved vascular oxidative stress, normalized aortic stiffness, and activated vascular SIRT1 activity. This connected to the clinical trials that were previously mentioned, as they hinted at the possibility and effectiveness of NMN in improving the cardiovascular system [40].

In many studies, NMN demonstrated the possibility of being developed for the benefit of patients with Alzheimer's disease.

This disease is known for its pattern of cognitive deficits and memory impairment. In one particular study, it is suggested that NAD+ may be essential to brain metabolism and boost memory and learning [40]. It relates NMN to boosting levels of NAD+, as shown in mentions of clinical trials earlier. The trial was held with mice, receiving NMN as treatment every other day for 28 days [40]. Results have proved that NMN treatment repairs cognitive impairments in year-old AD-Tg (Alzheimer's disease transgenic) mice that had impaired memory and spatial learning [40]. In the short period of NMN treatment, the mice demonstrated great improvement in spatial learning.

Adding to that, NMN displays great potential for possibly relieving age-associated susceptibility to acute kidney injury (AKI). It has been shown in experiments hosted on 20-month-old mice and proves to boost NAD+ and SIRT1 levels in aged kidneys [40]. SIRT1 is a type of renal protective molecule, sirtuin 1. At the same time, the supplementation of NMN is seen to protect these aged mice from cisplatin-induced AKI [40]. Cisplatin is a medication known for causing kidney damage as one of its side effects. SIRT1 is a key mediator in the aging of kidneys, and manipulating SIRT1 activities with NMN could potentially be a solution for AKI in elderly patients. Similarly, it would lead to more effective and specific treatments for AKI in relatively older populations.

Not only does the NMN have an influence on the organs, it also has an effect on the skeletal system. It has been discovered that NMN promotes MSCs within the living organism and in a laboratory setting. The MSC is a type of stem cell that can differentiate into multiple cell types, including bone and fat cells. One study on mice suggested that the SIRT1 protein was essential for NMN regulation of balance between bone and fat [40]. It is also implied that there is a potential connection between NMN treatment and its ability to address problems relating to osteoporosis in aged mice. Osteoporosis is a bone disease that develops when bone marrow density decreases, causing the bone to become weak. This is related to MSC, suggesting the possibility as NMN appears to influence positively [40]. NMN is also able to help restore balance between bone and fat cell formation in aged mice, suggesting the potential for NMN to become a therapy for combating the bone loss that comes with aging.

Possibly, NMN could also be developed to restore fertility that slowly deteriorates through female reproductive aging. In one experiment with 8-month-old mice, it was discovered that the replenishment of NAD+ would improve oocyte quality as well as increase ovulation rates and birth rates [41]. Age-related oocyte quality decline is associated with chromosome segregation defects. This might lead to abnormal chromosome numbers in embryos (aneuploidy). It is suggested that NAD+ deficiency could affect energy metabolism and cause aneuploidy. In this study, it was discovered that a lower dose of NMN produced an increased improvement in the functional fertility of aged animals, demonstrating the potential optimal range for these NAD+ precursors [41]. NMN treatment was also linked to changes in ovulation rate, potentially from decreased atresia of follicles during maturation [41]. Additionally, it improves embryonic development and stresses the potential role NAD+ may have in addressing age-related deficits in oocyte quality.

Overall, NMN is a supplement that has great possibilities and is seen to have many directions for development. It has not only been shown to assist pre-diabetic patients but also help patients with Alzheimer's disease and even acute kidney injury that comes along with aging. As a precursor of NAD+. NMN has also demonstrated safety in humans through the clinical trials often held upon aged adults, in which all clinical trials suggested some form of positive feedback on the intake of NMN.

CHAPTER 2

VEGANISM AND CARDIOVASCULAR DISEASE

Part 1: What is Cardiovascular Disease?

According to statistics, cardiovascular disease kills one person every 33 seconds in the US. Heart disease alone cost the United States about $239.9 billion each year from 2018 to 2019. Other examples of cardiovascular diseases include stroke and peripheral arterial disease. But to learn about how to prevent such dangerous conditions, you first must understand Cardiovascular disease, which leads you to the circulatory system. Essentially, the heart uses the pulmonary circulation to replace deoxygenated blood into the veins oxygenated blood. Then the oxygenated blood will be pumped from the heart to the rest of the body through systemic circulation. The exchange of oxygen molecules of the blood into the capillary tissues are then delivered to cells for essential processes that keep us alive. Secondly, you must understand the relationship of flow rate, pressure gradient, and resistance. This is a simple process that can be represented by an equation Flow= (Change in pressure)/Resistance. Where flow (the volume of blood passing through per unit of time) is directly proportional to the pressure gradient and inversely proportional to the vascular resistance . Therefore, when vascular resistance increases, the volume of blood flowing through per unit of time decreases and there is not enough oxygen exchange to support the circulatory system [1].

During the heart attack, your heart does not receive enough oxygen. This connects to the next point, what are some factors that increase resistance is blood viscosity, vessel length, and vessel radius. Blood viscosity means the friction between molecules of a fluid during a flow. This means factors commonly found in cardiovascular disease patients such as high LDL cholesterol increases blood viscosity, causing resistance to blood flow [1]. A build up of these fatty deposits can develop on the walls of coronary arteries, the vessels that supply oxygenated blood to the heart. This decreases the vessel radius and thus also increases resistance, and cells do not receive oxygen as effectively. In more serious cases, if one of these plaques ruptures, a blood clot will form which can fully close the artery and blood flow is cut off to the cardiac muscle and these cells to die. This is known as a myocardial infarction or more commonly known as heart attack. This is the causation for most cardiovascular diseases. For example, stroke is caused by interruption in the brain's blood supply, and peripheral artery disease is blockage of vessels to the peripheral systems [2]. Now that you have a better understanding of how cardiovascular disease occurs, how does veganism help prevent cardiovascular disease?

Cardiovascular disease can be caused by both clogged arteries cutting off blood flow causing oxygen deprivation and SCAD where there is a rupture causing slowing or blocking of blood flow. Therefore, in theory, a healthy plant related diet can help facilitate cholesterol and blood pressure associated with cardiovascular disease. For example, consuming too much sodium often found in pizza, cheese, chicken which will lead to high blood pressure, while fiber in whole plant foods can help ward off stroke by reducing artery-clogging plaque in blood vessels and lowering blood pressure. Thus, a hypothesis is formed based on the fact that a plant-based diet will lower the risk of stroke compared to other diets because vegetarians receive much fiber, potassium, and antioxidants that lower cholesterol and blood pressure associated with preventing artery clogging and bleeding that leads to cardiovascular disease [4]

The research paper titled, *Risks of ischaemic heart disease and stroke in meat eaters, fish eaters, and vegetarians over 18 years of follow-up: results from the prospective EPIC-Oxford study* (2019) by Tong T Y N, Appleby P N, Bradbury K E, Perez-Cornago A, Travis R C, Clarke R which studies the cases of ischaemic heart disease into three distinct diet groups, meat, fish, and vegetarians through online questionnaires over 18 years of follow up questions focusing on diet, lifestyle, health characteristics, and medical history which assessed diet groups. In addition, questions on socio demographic status, lifestyle, and medical history, such as physical activity, use of supplements, use of alcohol and smoking attributed to adjustment of data along with self reported cases of high blood pressure and high cholesterol levels. Overall, there were 2820 cases of ischaemic heart disease including heart attack from a total sum of 48188 participants [6]. Whereas 2324/31934 (.0728%) cases occurred in meat eaters while 496/16254 (.0305%) cases occurred in vegetarians containing vegans thus supporting the hypothesis that veganism is lower rates of ischaemic heart disease likely explained by “lower body mass index and lower rates of high blood pressure, high blood cholesterol, diabetes associated with these diets.”

5 studies were combined comparing vegetarians with non vegetarians with similar lifestyles, vegetarians are a group who did not eat fish nor meat. Mortality from ischemic heart disease was 24% lower in vegetarians than in nonvegetarians mostly observed in those at younger ages and was restricted to those who had

followed a vegetarian diet for more than 5 years. Supporting that dedication in a healthy vegetarian diet supports resistance against cardiovascular disease [2]. Things taken into account include body mass index, alcohol users, education levels, rate of exercise, and persons of old age >90 are taken out of the data. Although the difference between education levels between meat eaters and vegetarians, "Vegetarians had a lower mean body mass index, a lower percentage of vegetarians were current alcohol users, and a consistently higher percentage of vegetarians were high exercisers" all of these must also be taken into consideration when presenting said data. Still, the death ratio for ischemic heart disease for meat eaters is 72/42276 (0.0017) 33/24018 (0.0014) [5]. A possible explanation for the lower mortality among vegetarians is that they have lower serum total cholesterol concentrations than nonvegetarians. "Measurements of serum cholesterol concentrations in 3 of the studies showed lower total serum cholesterol concentrations in the vegetarians than in the nonvegetarians: 0.61 mmol/L lower in the Health Food Shoppers Study, 0.43 mmol/L lower in the Oxford Vegetarian Study, and 0.33 mmol/L lower in the Heidelberg Study [3]. Similar differences were observed in other cross-sectional studies comparing vegetarians with nonvegetarians. It has been predicted that a 0.6 mmol/L reduction in the total serum cholesterol concentration would cause a 27% difference in mortality from ischemic heart disease." Still, it is also possible that the lower mortality in vegetarians was due to other factors such as reduced oxidation of LDL cholesterol or changes in blood clotting [5].

LDL cholesterol (low-density lipoprotein), also called "bad cholesterol", makes up most of the cholesterol in human body. However, high levels risk stroke and heart disease, the most potent source of this type of cholesterol include bakery items, full-fat dairy products, red meats, processed meat, and fried foods [6].

Part 2: Types of Diets, Pros and Cons

A vegetarian diet excludes all meat. However, some choose to eat only fish as a source of meat – pescetarianism. Some benefits of adopting a vegetarian diet include reduced risk of heart disease, cancer, and joint pain caused by arthritis. This is maintained by having a well-balanced diet of grains, vegetables, and dairy. On the other hand, vegetarians lack iron in their body and tend to have greater risk of iron-deficient amnesia. Eating beans, quinoa, nuts, tofu, and lentil are vegetarian sources of iron. Vitamins and supplements may also be necessary depending on the individual's state of health. Similar to a vegetarian diet, a vegan diet solely focuses on plants and excludes food from animals or any animal products including dairy.

The Paleo diet is commonly referred to as the accustomed diet prior to the age of modern farming. Diary, legumes, refined sugar, and salt are excluded. The Paleo diet includes lean meat, fish, fruits, and vegetables. It is linked to weight loss, better blood pressure, increased glucose tolerance, and lower triglycerides. Because the Paleo diet excludes legumes, grains, and dairy, this often causes deficiency and lack of nutrients in fiber, vitamins, and calcium.

Often known as the diet originally used to reduce seizures in children with epilepsy, the ketogenic diet consists of a low carbohydrate, high fat intake. The keto diet attempts to force the body to use ketone bodies which is a type of fuel used from stored fat instead of carbohydrates. The calories are replenished in fats in meat, eggs, and fish. Hence, it is reasonable that this diet is correlated to effective weight loss, reduced risk of heart disease and cancer, and decreased rates of LDL cholesterol. A keto diet is associated with reduced visceral adiposity, a hormonally active component of total body fat. The diet trains the human body to burn fat as fuel rather than carbohydrates. Due to not eating a wide range of vegetables and fruits can cause deficiency in including selenium, magnesium, phosphorus, and vitamins B and C. Lack of dietary fiber from this diet can also cause constipation, and liver problems since due to too much fat to metabolize.

The "heart-healthy diet" or more recognized as the Mediterranean diet, daily consumption of fruits and vegetables, weekly intake of fish, moderate portions of dairy, and limited amounts of red meat. The Mediterranean diet is linked with reduced risk of Alzheimer's, Parkinson, and type 2 diabetes as well as slower decline in brain function, lower risk of certain types of cancer, and a healthier digestive system. However, because of a lack of protein, there are often calcium and iron deficiencies as a result of adopting this type of diet.

The information above illustrates the different kinds of diets, each with their advantages as well as disadvantages. This section of the chapter will now dive deeper into the reasons why a vegan diet has by far, had some of the most extensive health benefits to offer.

According to the American Diet Association, 'appropriately... planned vegan diets.. are appropriate for individuals during all stages of the life cycle' (ADA, 1266). Concerns have been raised over the years of lack of protein, vitamin B12, essential fatty acids, iron, and calcium associated with a vegan diet. However, the variety of vegan food options readily available and supplements make up for these deficiencies. For protein, peas, lentils, and beans are good sources of protein in addition to 'cereal-based diets, especially those based on wheat and maize, [which] supply protein levels considerably above the requirement level' (Millward, 259). Hence, protein can be obtained through a range of plant foods.

Well-planned vegan diets which include abundant fruits and vegetables contain large amounts of magnesium and potassium while nuts and grains increase a high renal acid load, the rate of intestinal absorption. Leafy green vegetables are low in oxalate and high in both calcium and vitamin K. Adequate amounts of calcium and vitamin D are equivalent to good bone health.

Although no plants are capable of producing vitamin B12, those who eat plant foods inadvertently gain adequate amounts of the vitamin. Growing concerns of B12 deficiency include being a part of risk factor for neurological disorders and cardiovascular diseases including pernicious anemia, decreased abnormal production of red blood cells. However, the Nutritional Society shows that vegan diets are more beneficial for people of older age as vitamin B12 binds to the protein in animal foods, impending smooth food absorption, equaling to lack of complete nutrients.

Lack of Vitamin D is often associated with bone problems and other conditions such as multiple sclerosis, hypertension, and cancer.

Part 3: Analysis of Case Studies and Diet Recommendations

When Rebecca McCormack picked up 35 year old Jennifer Valentine for a 4.5-hour flight. Contrasting her usual eating routine, Valentine ordered fried hash browns and a gravy biscuit when she had just begun exercising and eating more nutritious meals two months earlier, recently losing 12 pounds from a high of 229. On the plane, everything was peaceful when Valentine's breath became labored and she couldn't understand why. She told McCormack that she felt fleeting pain, became nauseous and then overheated. And began to shake so badly she couldn't hold the orange juice she'd ordered. Still Valentine endured the flight and on the way to the terminal, Valentine became incredibly thirsty and began to cry.

At the hotel, Valentine went to take a nap. She called her husband to let him know what happened. She got up to comb her hair and once again, she felt out of breath. She went to McCormack's room in tears. They called a doctor and she was sent to urgent care, where a nurse performed an electrocardiogram. She tore the results off the machine and left the room. When she returned a few minutes later, she told Valentine: "An ambulance is on the way. You need to go to the ER."

An ultrasound and an echocardiogram showed no issues. She was then taken to the cardiac catheterization lab for further diagnostic tests. An hour later, when they woke her up, a doctor told her she'd had a heart attack. He showed her two images. One was of her right coronary artery, 100% blocked. Blood wasn't flowing. The other was after they placed a stent. Blood was flowing.

Doctors told Valentine that her heart attack was from a plaque rupture and that the cause was genetic. They also said she had a 65% blockage in her main coronary artery, but they would not place a stent until it became further blocked. Valentine had not concerned herself with her family history of heart issues until this day. Claiming that "When you're younger, you don't think it will happen to you."

Valentine was determined to continue to lose weight and exercise. Although she couldn't change her genetics, she knew that becoming healthier would reduce her risk of heart disease. Still Valentine stated that "I felt like I was a ticking time bomb, anytime I was trying to plan groceries, I'd just start crying. I didn't even know what I could eat." Anxious about having another heart attack, she started to have panic attacks and took medication for anxiety and depression.

She spoke with a nutritionist and started following recommendations from the American Heart Association. She restricted her sodium and saturated fat intake. She changed her bread and flour from white to whole wheat. She cut out most processed foods. She also started walking daily, and in the years since the heart attack, which happened in 2018, Valentine has lost 92 pounds. She now weighs 137 and has stopped taking anxiety medication. Now 40, Valentine shares her achievements online to encourage others to eat better and exercise more.

She also inspired her friend, McCormack, to exercise and practice a healthy diet with her. "I was a terrible eater, eating for convenience instead of for my body," said McCormack, who also has heart disease in her family. "Watching Jennifer helped me learn to make better choices." Each helps the other strive toward a longer, healthier life.

A 68-year-old man approaches the pharmacy counter looking for advice on his mild mitral valve regurgitation, which is a condition where the valve between the left heart chambers doesn't close completely, allowing blood to leak backward across the valve. Mitral valve regurgitation causes issues where less blood will move through the heart and the rest of the body. As a result, it causes fatigue or shortness of breath. He is unsure what self-care measures are necessary for this condition. Upon questioning, he states that he is a cigarette smoker and has a history of hypertension (high blood pressure) and hyperlipidemia (high cholesterol), managed on a combination of amlodipine, valsartan (drugs used to lower blood pressure via relaxing blood vessels), hydrochlorothiazide (a drug used to control blood pressure through treating fluid retention), and atorvastatin (a drug used to slow the production of cholesterol in the body). Pharmacists advise that he consider quitting smoking since smoking not only increases risk of heart attack and stroke but may also worsen his mitral valve disease. They also advise him to understand the importance of a healthy diet, such as consideration of the DASH (Dietary Approaches to Stop Hypertension) diet which mainly consists of foods rich in minerals potassium, calcium, and magnesium thus focuses on food like fruits, whole grain, and vegetables while also including for fat free and low dairy products. The DASH diet also limits saturated fat, sodium (salt), and alcohol consumption. This diet is ideal for optimizing blood pressure control and heart health by lowering cholesterol, along with routine exercise and physical activity for managing his heart function.

A-55 year old female patient was referred to a clinical nutritionist to start dietary changes. The patient had some key cardiovascular disease risk factors including hypercholesterolaemia (higher cholesterol levels at birth), history of smoking, hypertension, and a family history of cardiovascular disease with her father dying at the age of 45 due to acute myocardial infarction (heart attack). The clinical nutritionist evaluated her diet, using a 3-day dietary recall of her eating habits. Although the result showed that the cholesterol and fat intake fits the recommended amount, several factors stood out: 1) the patient was consuming too much saturated fatty acid per day mainly from consuming butter and cheese and 2) the patient was consuming too much fiber per day. Therefore, a clinical nutritionist recommended replacing butter with vegetable oil based margarine. This is because butter is made from cream or milk which is mostly saturated fat, meanwhile margarine is a blend of oils that are mostly unsaturated fat. Fat in plant oils lowers total cholesterol and low-density lipoprotein cholesterol, or "bad" cholesterol, when it takes the place of saturated fat. On a molecular level, saturated fat has hydrocarbon chains connected by single bonds only, while unsaturated fat includes one or more double bonds in its hydrocarbon chains, with one fewer hydrogen atom on each double bonded carbon. This creates a kink in the chain and makes unsaturated fat easier to be broken down by your body, while saturated fats can cause plaques to develop within the walls of blood vessels and contribute to cardiovascular disease.

Additional recommendations for diet change included using vegetable oils in cooking and adding them to salad dressing and replacing white bread with wholemeal bread. after two months of follow up, the the patient's serum and LDL cholesterol values had lowered from a serum total cholesterol of 228 mg/dl to 216 mg/dl (5% decrease) and LDL-cholesterol of 174 mg/dl to 154 mg/dl (11% decrease). However, these changes were not sufficient since an LDL-cholesterol goal of less than 100 mg/dl is ideal in respect of the patient's moderate risk for developing cardiovascular disease. However, the patient still denied drug therapy. Therefore, the patient was advised to change their vegetable-oil based margarine to plant sterol/plant stanol containing margarine so that 2 g of plant sterols or stanols were consumed daily. These substances can be seen as cholesterols made in plants and helps limit the entry of cholesterol into the body and in some plant sterols, it also reduces the amount of cholesterol made in the body. After another 2 months, the patient's serum total cholesterol concentration was 194 mg/dl (-34 mg/dl, 15% reduction) and LDL-cholesterol concentration was 135 mg/dl (-39 mg/dl, 22% reduction). Although the change was very significant, it still did not reach the target goal of LDL-cholesterol concentration being less than 100 mg/dl. Therefore, combining statin therapy with additional dietary changes was considered.

Although drug use in the end was inevitable in this case, due to implementing statin usage at a much lower cholesterol base point than previously, a much smaller dose is needed than what the patient would need before the previously implemented dietary changes and the patient having a much higher risk of experiencing statin side effects such as headaches, nausea, muscle pain, and joint aches. Or in some scenarios, more serious conditions such as diabetes, muscle cell damage, memory problems, and liver damage.

A 37-year-old man undergoes a neurological examination after complaints of numbness and sensation of warmth combined with right-sided hemiparesis (the inability to move one side of the body). The patient had also reported a rash on the abdomen and right elbow; he had been in urgent care and started on a prednisone taper. (Prednisone is a drug that decreases inflammation and relieves allergic reactions.) He was also diagnosed with colitis (inflammation of the colon's inner lining). As a result, he is now taking ciprofloxacin (used to treat or prevent certain infections caused by bacteria, such as pneumonia) and metronidazole (stopping the growth of certain bacteria and parasites).

The patient also had a history of chronic essential hypertension, hypothyroidism (thyroid gland doesn't make enough hormones, causes conditions such as fatigue, weight gain, slowed heart rate), and Hodgkin lymphoma (cancer affecting immune system) of left side of neck without metastasis (further growth away from primary site) at the age of 10; and sleeve gastrectomy (a surgical weight loss procedure) with bilateral ischaemic strokes one year ago with residual left arm and leg weakness. He had a 180 lb weight loss after surgery. Since the stroke, the patient was on 81 mg of aspirin daily and 40 mg of atorvastatin daily. A head CT was performed, and three chronic infarcts were discovered in the right MCA (damage to the right of the middle cerebral artery, thus explaining the deficit in the left side of the body) and chronic infarct involving the left medial occipital cortex. A brain MRI then revealed areas of encephalomalacia (cerebral softening, bleeding, and inflammation causes softening of brain substance) and gliosis (proliferation of glial cells in injured regions of the central nervous system). These were located in the inferolateral right frontal lobe, right posterior frontal lobe, right parietal lobe, and medial left occipital lobe, relating to his previous cerebrovascular accident.

In addition, transthoracic echocardiography revealed mild aortic stenosis (narrowing of blood vessels) with mitral valve calcification (excess calcium). Transesophageal echocardiography (TEE) found a complex atheroma measuring 9 mm thickness near the sinotubular junction in the ascending aorta. Treatment for the patient included rivaroxaban (to treat and prevent harmful blood clots) and rosuvastatin (lower LDL, fats, and raise HDL) for secondary prevention of cardioembolic stroke after discovering the complex atheroma. In addition, aspirin was given to the patient, and symptoms resolved in a couple of hours. After being readmitted two months later and no acute stroke discovered, the patient was discharged on aspirin 81 mg daily, rivaroxaban 20 mg daily, and rosuvastatin 5 mg daily.

The exact cause of his stroke-like symptoms was unclear. Possibly due to the reason include cardiogenic ischaemic stroke, hypercoagulable state from lymphoma, or aortic stenosis secondary to radiation therapy. Stroke is the fifth leading cause of death in the US, but young

adults only account for 10-15% of all stroke cases. However, this percentage is rising; the most common risk factors include smoking, dyslipidemia, and hypertension, one of which the patient does exhibit. These conditions can all be controlled, smoking can be prevented, and studies and case reports have shown that dyslipidemia and hypertension can be controlled with a veganism-related diet.

More severe examples of cardiovascular diseases include a 67-year-old woman seeking emergency medical care due to prolonged chest pain where she was diagnosed with myocardial infarction, a condition where there is a lack of blood flowing into the heart. The patient had a history of hypertension, diabetes mellitus, dyslipidemia and smoking. X-ray screening revealed lesions with 70% obstruction in the right coronary, anterior descending and circumflex arteries. A left ventriculography revealed apical akinesis with signs of intracavitary thrombus in that region. After being recommended for drug and clinical treatment, the symptoms subsided until 5 months later, where she experienced a lack of blood flow to the brain and severe chest pain. In December of 2009, after admission to the hospital, the patient experienced 3 cardiac arrests with the final one being irreversible and the death of the patient. With the final hypothesis for the reason of death being acute myocardial infarction complicated by cardiogenic shock.

Eight years ago, a woman in her 50s was diagnosed with idiopathic (unknown cause, peculiar) pulmonary arterial hypertension, understood as high blood pressure in the lungs. The patient experienced symptoms such as shortness of breath, fatigue, dizziness, and palpitations. The resting mean pulmonary artery pressure (abbreviated to mPAP with the standard range being 14.0 ± 3.3mmHg) was 48 mmHg, pulmonary vascular resistance of 10 WU (pulmonary vascular resistance is defined as the resistance against blood flow from the pulmonary artery to the left atrium, abbreviated to PVR. Standard range being around 2.1 WU), cardiac index of 2.3 L/min/m^2 (Cardiac output or amount of blood pumped into the systemic circulation per unit of time. The cardiac index is cardiac output per square meter of body surface area; the standard cardiac index for adults is 3L/min/m^2), and a pulmonary capillary wedge pressure is 9 mm Hg (estimates left atrial pressure). Specific PAH therapy was used, such as tadalafil, macitentan, and treprostinil IV (Chrono five infusion pump device), supplemented with warfarin, spironolactone, and long-term oxygen therapy (2 L O2/min). However, even with the more intensified treatments, the PAH symptoms progressed from a World Health Organization functional class of III to IV, and symptoms worsened. (WHO FC of III is defined as normal activities around the house being greatly limited from shortness of breath, and WHO FC of IV being symptoms both at rest and severe with activity.)

Four years ago, the patient was listed for bilateral lung transplantation. An electrocardiogram revealed right bundle branch block and placement of an implantable loop recorder for cardiac monitoring to assess arrhythmias, heart rate variability (HRV) and heart rate. Surprisingly, no arrhythmias were observed in the 100-day period pretransplantation. Post-BLTX, the patient experienced clinical improvement and CMR showed complete normalization of cardiac index, EDV, SV, left and right ejection fraction. This is an example of a more severe case of pulmonary arterial hypertension of WHO FC level of 4 where symptoms interfere with daily life functions leading to repeated hospitalizations. However, pulmonary hypertension develops slowly and many have it for years before symptoms grow too severe. This can be controlled easily by making simple lifestyle changes such as a healthy vegetable related diet to control factors such as blood pressure to control development of cardiovascular disease such as these.

When compared with people who engaged in a plant-based diet, non-vegetarians tend to have higher body mass index, LDL-cholesterol, and unhealthy overall lifestyles (alcohol drinking, smoking, poor nutritional values in food). Risk factors such as high cholesterol stem from eating animal products and processed foods, both of which plant-based diets discourage. [7]

In a case study done in India, results had proved that a nutrition transition from a vegetarian to a non-vegetarian diet contributed to a negative correlation to cardiovascular diseases such as stroke, weight gain, and noncommunicable diseases. India, which has a long-standing tradition of vegetarianism, emphasizes brown rice, roots, and pulses. Use of unsaturated fats is also promoted. However, when vegetarians switched to a diet consisting more of meat protein, it increased the risk of stroke to 152% and decreased life expectancy down 3.6 years. In addition, it increased the likelihood of weight gain, raising the BMI. The change of diets shows an evident difference in particularly potent risk factors. Stroke is a type of cardiovascular disease that has a profound impact on the daily lives of individuals. When those who committed to a vegan diet were tested, it was shown by data of the many health benefits that were a result of this consistency.

In the study case done in Taiwan, when cases of each type of stroke were separated, vegetarians in cohort 1 experienced lower risk of ischemic stroke and in cohort 2, they experienced lower risks of stroke. Thus, data strongly suggests that many types of cardiovascular diseases such as stroke and its complications heavily rely on consumption of food and other products surrounding veganism. Plants contain major antioxidants and dietary fiber. [8]

The EPIC-Oxford study proved that a healthy plant-based diet was associated with over 10% lower risk for total stroke and marginally lower rates for ischemic stroke. A healthy plant-based index largely contributed to the health outcomes of individuals. While some calculations determined that there was an inverse association between vegetarians and stroke rates, those that were based on a plant-based diet but consumed unhealthy plant floods (refined sugars and grains) influenced the data results. [9]

Veganism largely plays a role in the likelihood of being diagnosed with any type of cardiovascular disease and aids in preventing these occurrences from happening. In addition, those who consume based on a vegan diet are less likely to smoke and have more consistency in other habits, such as going outside and consuming less amounts of alcohol. Starting a vegan diet and committing to these habits prove to be beneficial to the physical body.

A 79-year old man who previously was able to walk more than one and a half miles began to develop serious health problems involving three-vessel coronary heart disease, moderate to severe aortic regurgitation, and left ventricular systolic dysfunction. Contrary to what was suggested, a valve replacement surgery and coronary artery bypass grafting, he followed a whole-food plant based diet to avoid surgery. Health records collected are as follows: his BMI was 26.6 kg/m^2; blood pressure was 127/50 mmHg while heart rate was 49 beats/min. Cholesterol wise, his total was 201 mg/dL and his lipoprotein cholesterol was 105 mg/dL.

After adopting a plant-based diet which consists of vegetables, whole-grains, legumes, and nuts excluding animal-derived products such as eggs, dairy, and meat, his overall health and wellbeing improved as demonstrated by his medical follow-ups. His BMI changed to 23.9 kg/m^2, blood pressure was 129/50 mmHg, and total cholesterol fell to 137 mg/dL while lipoprotein cholesterol dropped to 51 mg/dL. It is important to note that during this time period, no cholesterol lowering medications or supplements were used. The changes in these levels were dependent on his diet, exercise regime, and daily routine.

Medications that he consumed before these changes in dietary patterns consisted of aspirin, atenolol and candesartan. Atenolol and candesartan are drugs used to control high blood pressure. Even with these medications to prevent stroke and heart attack, he developed severe health problems. However, 2 months later after his conversion to a whole-food plant based diet, his prescription was as follows: atenolol, which was changed to carvedilol used to treat hypertension, pravastatin, and continued intake of aspirin and candesartan. Knowledge provided showed that his left ventricular ejection increased to 50%. This indicates the percentage of blood pumped out of the left ventricle every heartbeat. For individuals in their 70s, a fraction of 50% to 65% is the range of a healthy ejection fraction. Prior to this, he had lower than 50%, evidence of an evidence cause of coronary heart disease and failure in ventricular dysfunction. While other factors could have contributed to this favorable outcome, in the context of his minimal medication and changes in cholesterol, blood pressure, and BMI, most evidence shows that his adoption of a plant-based diet played a meaningful role in this change in results. [10]

A 54-year old female who prior to her change in diet, had grade 3 obesity and type II diabetes. An echocardiography test was used to show blood flow through the heart and valves, showing a 25% left ventricular ejection fraction, an indication of heart failure. Upon her diagnosis, she altered her lifestyle and diet patterns, adhering to a whole plant-based diet after her "healthy Western diet" proved of no positive results. Her diet regime consisted of no animal products, 1-3 servings of dark, leafy greens, fruits, whole grains, nuts, and limited processed/packaged food [11].

After 5 months since her commitment to a plant-based diet, health parameters showed the profound impacts these small, yet accumulative gains in improving health had on her. Prior to this, her BMI was 45.2 kg/m^2; hemoglobin Alc, a test to determine blood sugar levels in a period of 3 months showed a baseline of 8.1% and the patient's ejection fraction to be at 25%. Documented results collected after 6 months were analyzed. She lost over 22.7kg and hemoglobin Alc was significantly lower, to 5.7%, a 2.4% decrease without the use of any diabetes medication. Ejection fraction was raised to 55%, which follows the normal ejection fraction of people in their 50s. In addition to her plant-based diet, she also ate supplemental vitamin B12.

Through multiple studies, proposals and conclusions supported by recent data have been reached. Vegetarian diets compared to diets consisting of meat, dairy, and egg products are associated with lower rates of plasma levels of myeloperoxidase, a pre-oxidant compound. HIgh levels of myeloperoxidase is associated with increased risk of heart disease and promotes plaque formation, leading to possible ruptures and myocardial infarctions.

A vegetarian diet is also associated with lower systolic and diastolic with increase in potassium and decrease in sodium concentration. Hypertension rates were also lowered compared to those who consumed foods based on an omnivorous diet. Soluble dietary fiber is only found in plant-based foods which also consists of short-chain fatty acids which regulate the enterocyte's expression of genes, lowering serum cholesterol levels. High levels are often correlated to an increased risk of coronary artery disease and myocardial infarction where heart muscles are affected by lack of blood flow and the formation of blood clots.

CHAPTER 3

VEGANISM AND DIABETES

Part 1: Introduction to Type 2 Diabetes

Diabetes is a disease affecting millions of patients worldwide. Among the 37 million Americans with diabetes, 90~95% of them have Type 2 Diabetes. More children and young adults are now developing Type 2 Diabetes, despite the chronic disease mainly developing in people over 45 years old [1]. Thus, Type 2 Diabetes is a growing concern in the United States, other Western countries, as well as various Eastern countries.

The diabetes epidemic demonstrates how important it is to address the management strategies, complications, and risk factors of Type 2 Diabetes. Healthcare providers, policymakers, and the whole of society should pay attention to diabetes as it is a multifaceted condition.

With Type 2 Diabetes, the body is dysfunctional in regulating glucose as fuel. In the long term, this dysfunction causes excess sugar to remain in the bloodstream, leading to immune, circulatory, and nervous disorders [2]. Diabetes causes cells to display resistance against insulin, a hormone made by the pancreas that allows blood sugar to enter the cells as fuel. Insulin resistance results in the pancreas producing more insulin so that the cells respond, which increases blood sugar levels and leads to prediabetes, then Type 2 Diabetes. High blood sugar damages the body and can lead to vision loss, heart disease, and kidney disease.

In multiple Western countries such as Canada, the United Kingdom, Australia, and parts of Western Europe, Type 2 Diabetes cases have been rising. This trend is connected to lifestyle aspects such as poor dietary choices, a sedentary lifestyle, and the modern Western diet with aspects of high sugar and processed foods.

On the other hand, diabetes has become more prevalent in Eastern countries like China and India, as traditional diets and lifestyles are becoming Westernized. The diabetes epidemic in these regions is propelled by an adoption of processed foods, less physical activity, and urbanization. As more individuals are affected by Type 2 Diabetes worldwide, it becomes more crucial to address this health challenge extensively. This includes not only realizing the causes and the consequences but also promoting prevention, being diagnosed early, and managing diabetes effectively to lessen the impact of this widespread condition.

Diabetes is a threat to eye health. Prolonged high blood sugar levels can damage the small blood vessels in the eyes, leading to a condition known as diabetic retinopathy. Diabetic retinopathy is a leading cause of vision loss and blindness among adults. It can manifest as blurred vision, floaters, and, in advanced stages, severe visual impairment. Regular eye exams and diligent blood sugar control are crucial for preventing and managing this potentially devastating complication.

Kidney disease, also called diabetic nephropathy, is a grave condition caused by diabetes. The kidneys, which are vital in filtering waste and extra fluid from the bloodstream, can be damaged over time by high blood sugar levels. The kidney's delicate filtering units become impaired, causing the kidneys to not function properly and waste products to build up. In serious cases, end-stage renal disease (ESRD) can occur, meaning that dialysis and kidney transplants are necessary for survival. Nerve damage, also known as neuropathy, is an underestimated side effect of diabetes that can influence different parts of the body, causing tingling, numbness, and pain most commonly in the extremities. Also, uncontrolled diabetes can lead to autonomic neuropathy, influencing nerves that control vital functions such as heart rate, bladder function, and digestion. Neuropathy can be a burden by causing chronic discomfort and diminishing mobility.

An individual's immune system can also become more susceptible to infections if the individual has diabetes. If diabetes is poorly managed, minor wounds and infections can escalate into severe complications. Serious infections and limb amputation can develop from foot ulcers if not properly and quickly treated. The risk of skin, respiratory, and urinary tract infections increase with a compromised immune system due to diabetes. In addition, diabetes collides with mental health issues, increasing the complexity of dangers. Furthermore, blood sugar level fluctuation can impact mood and cognitive function, so it is essential to address the psychological components of diabetes to create a holistic approach to well-being.

As such, diabetes has many dangerous consequences that extend far beyond high blood sugar levels. Diabetes can result in a cascade of serious complications impacting the heart, nerves, eyes, mental health, and the immune system. The risks associated with diabetes highlight the necessity in preventing diabetes, managing the disease, and creating public awareness to avoid the life-altering and possibly life-threatening results.

Part 2: Solutions to Type 2 Diabetes – Author Research

As Type 2 Diabetes is a serious problem, this chapter attempts to address the major issue of the prominence of Type 2 Diabetes in modern-day society and provide sufficient information for readers to inform them of the potential method to reduce their risk of Type 2 Diabetes. Although several diabetes medications exist, such as Metformin (Fortamet, Glumetza) as well as insulin-inducing Sulfonylureas and pancreas-stimulating Glinides (to secrete more insulin), there exists a faster, more efficient, and much cheaper alternative. It is a possible preventative method that has the potential to decrease the risk of one getting Type 2 Diabetes: a plant-based diet.

Plant-based diets or plant-forward eating patterns focus on foods primarily from plants, including not only fruits and vegetables, but also nuts, seeds, oils, whole grains, legumes, and beans. Plant-based diets have also been shown to "offer all the necessary protein, fats, carbohydrates, vitamins, and minerals for optimal health, and are often higher in fiber and phytonutrients" (3). In addition, Plant-based diets have been demonstrated "to support [the] immune system, to reduce inflammation, to help maintain a healthy weight, and to contain high amounts of fiber, as well as to lower cancer risk" [4]. Another important recently discovered potential advantage of a plant-based diet is that it may be "especially beneficial with those with Type 2 Diabetes" [5].

Part 3: Case Studies

This chapter's research involves assessing the validity of the statement by compiling information from several original research papers comparing the effects of a plant-based diet to a non-plant-based diet, then calculating the rate of the development of Type 2 diabetes in subjects with a plant-based diet and a non-plant-based diet, followed by a chi-squared analysis to discern whether a plant-based diet truly has an effect on the rate of diabetes.

For this chapter's research, all those in groups not vegan are considered to partake in a non-plant-based diet due to the inclusion of poultry and fish in their diets. There are three primary articles that had data taken from to use in this chapter's calculations: *Type of Vegetarian Diet, Body Weight, and Prevalence of Type 2 Diabetes (2009)* by S. Tonstand, T. Butler, R. Yan, and G. Fraser, which focuses on the "assessing the prevalence of Type 2 Diabetes in people following different types of vegetarian diets compared with that in nonvegetarians" through a study with data coming from a "50-page self-administered questionnaire" that included sections on "illness, diet, physical activity, demographics, height, and weight". Participants in the study were from North America. Cases of diabetes were "ascertained by asking whether a physician had ever diagnosed Type 1 Diabetes or Type 2 Diabetes". Further "allocation of subjects into dietary categories...including meat, fish and poultry, dairy, and egg consumption" and the "telephone interviews with subjects" allowed researchers to determine the percentage of vegans with Type 2 Diabetes, (2.9% of 2,731), and the percentage of non-vegans with Type 2 Diabetes (21.7% of 58,172) after adjustment for age, sex, ethnicity, education, income, physical activity, television watching, sleep habits, alcohol use, and BMI.

Likewise, In *Vegetarian diets and incidence of diabetes in the Adventist Health Study-2 (2013)* by S. Tonstad, K. Stewart, K. Oda, M. Batech, R.P Herring, and G.E Fraser, researchers grouped participants (coming from both the United States as well as Canada) as "vegan, lacto-ovo vegetarian, pesco-vegetarian, semi-vegetarian, or nonvegetarian." After grouping participants, a "follow-up questionnaire after two years elicited information on the development of diabetes" was given to participants. After the experiment was controlled for age, gender, education, income, television watching, physical activity, sleep, alcohol use, smoking, and BMI, it was determined that vegans had a lower risk of diabetes than non-vegetarians (0.54% of 3,545), lacto-ovo vegetarians, pesco-vegetarians, and semi-vegetarians (combined total of 5.42% of 37,842).

For the third research study titled *Vegetarian diet change in dietary patterns, and diabetes risk: a prospective study* (2018) by T. Chiu, W. Pan, M. Lin, and C. Lin, researchers investigated the "association between a vegetarian diet, change in dietary patterns and diabetes risk in a Taiwanese Buddhist population" by following a population of "non-smoking, non-alcohol drinking Buddhists free of diabetes, cancer, and cardiovascular diseases at baseline, for a median of 5 years". Diet was also "assessed through a validated food frequency questionnaire at baseline and a simple questionnaire during follow-ups". Similarly, the incidence of diabetes was discerned through "follow-up questionnaires, fasting glucose, and HbA1C (average blood sugar level test)". At the end of the experiment, it was determined that for the group with a plant-based diet, there was a lower risk of diabetes (1.01% for 5431) whereas the non-plant-based diet group had a 1.7% chance of diabetes for a 5,456 group of participants after adjusting for age, sex, education, and a family history of diabetes.

As such, these three studies were chosen due to the clear distinction between a plant-based control group and a non-plant-based control group for the experiments as well as the adjustments of factors for a more accurate result. The data from the experiments were used for a meta-analysis.

For the chapter's analysis on possible effects of a plant-based diet on Type 2 Diabetes, both online and physical materials were used. Resources used were the Google Search Engine, Google Spreadsheets for quantifying data, Google Documents for recording data, and Google Drive for storing analysis and data. In addition, a Ti-84 calculator was used to conduct chi-squared analysis. Furthermore, the chapter's analysis referred to online resources such as research papers, online studies and organizations, diabetes webpages that discuss insight on diabetes and the role of specific diets, as well as the National Institutes of Health (NIH) Database to look at articles. Other online resources included studies that explained the benefits of a plant-based diet, meta-analyses, and a chi-squared value to p-value calculator from Social Science Statistics (https://www.socscistatistics.com/pvalues/chidistribution.aspx).

Overall, the method of the analysis utilized the Google Search Engine to search for and take original research papers focusing the effects of a plant-based diet on Type 2 Diabetes. These research papers were carefully chosen based on specific criteria, such as whether subjects were divided into plant-based or non-plant-based groups, as well as the data on the number of subjects who had Type 2 Diabetes within the research period. Extensive notes were taken of the research papers, and each article was studied carefully to take out the most important information.

Afterwards, the speed of subjects that developed Type 2 Diabetes in the plant-based and non-plant-based groups were calculated. This was done by dividing the amount of individuals with Type 2 diabetes by the total number of subjects in each group. The relative change formula was used to find the difference in Type 2 diabetes rates among the two groups.

To calculate the statistical significance of the data, a null hypothesis was created: the rates of Type 2 diabetes in the plant-based and non-plant-based groups have no statistical difference. A chi-squared analysis was conducted using the observed values.

Using a p-value converter, the chi-squared value was converted to a p-value at a degree of freedom of 1 and significance level of 0.05. The results are statistically significant if the p-value is less than 0.05, meaning that the null hypothesis would be rejected and that there is a difference in the rates of Type 2 diabetes between the plant-based and non-plant-based group. In comparison, if the p-value is greater than 0.05, it means that the results are not statistically significant and that the null hypothesis is accepted. Thus, there would be no difference in the rates of Type 2 diabetes between the plant-based and non-plant-based groups.

A culmination of the data gathered by several research articles that distinguish between plant-based and non-plant-based diet groups for the experiments of researchers was done. Utilizing the data found from the graphs, a total plant-based diet diabetes rate was able to be determined by adding up the rate of diabetes from each study for each category (plant vs non-plant). Diabetes rates calculated were 2.1% and 4.1% for plant-based and non-plant-based, respectively, calculated by putting the total number of people with diabetes over the entire number of people in select groups (either plant-based or non-plant-based).

From here, values calculated for the relative rate or difference in diabetes rate were utilized. By utilizing the relative rate formula and plugging in 2.1% as the initial and 4.1% as the final value, a 95% difference in diabetes rate of a non-plant-based diet from a plant-based diet is yielded, showing that there is quite a significant difference in diabetes rate between the two groups.

However, in order to determine whether there is a significant difference between observed data and expected data, or if it is simply the result of chance, a chi-squared test was utilized to ensure that there is indeed a relationship between the variables tested. In order to attain the expected values, however, the overall total rate of diabetes for the entire meta-analysis was calculated. This was done by adding up all the participants across all three studies, yielding a total population size of 104,387 participants. The total number of participants who ended up with Type 2 Diabetes was also calculated in the same manner, yielding a total of 4,158. Again, only two variables for chi-squared are used (only those with diabetes from plant-based and non-plant-based groups because this chapter's research project focuses on the decrease in the risk of diabetes that a plant-based-diet confers, rather than the increase in the probability of being non-diabetic (using four variables for chi-squared: diabetes from plant-based and non-plant based as well as non-diabetes from plant-based and non-plant-based). Using four variables would conflate the two different questions. Dividing 4,158 by 104,387 yielded 3.98%, or the overall rate of diabetes in the study. As such, the expected was calculated by multiplying the 3.98% value with the total number of plant-based diet participants (7329), and non-plant-based participants (97058), resulting in the expected values of 292 and 3866 (rounded), for the respective groups. Plugging these values into the chi-squared equation yielded a $\chi 2$ value of 71.17. After plugging $\chi 2=71.17$ into a chi-squared to p-value converter in order to determine whether the null hypothesis should be rejected or kept, a value of $p<0.0001$ is yielded. As $p<0.05$, the null hypothesis can be rejected.

Part 4: Conclusion and Recommendations

The data gathered from the series of experiments involving large groups of people for both the plant-based and non-plant-based groups and following a chi-squared analysis with two variables (diabetes in plant-based diet and diabetes in non-plant-based diet) that resulted in a p-value<0.0001 ($p<0.05$), demonstrates the fact that (due to the rejection of the null hypothesis; null hypothesis: there is NO statistical difference between the rate of diabetes of a plant-based diet group and the non-plant based diet group) there is indeed a statistical difference between the rate of diabetes of a plant-based diet group and a non-plant based diet group. This indicates that a plant-based diet does indeed have an effect on diabetes.

There are many reasons explaining why exactly a plant-based diet contributes to a lower risk of diabetes. For instance, a diet composed of plant-based foods contributes to lower insulin resistance, (when cells stop responding to the insulin that is being pumped by the pancreas) a hallmark of Type 2 Diabetes. Plant-based diets, however, improve insulin sensitivity in cells rather than insulin resistance, as they are high in "fiber, antioxidants, and magnesium, all of which have been shown to promote insulin sensitivity" (6), thus lowering the risk of diabetes for those that have plant-based diets.

Furthermore, plant-based diets that include whole grains, fiber, vegetables, and nuts are high in fiber, which all have been found to be protecting against diabetes (7). It was also noted that, specifically, whole grains, including whole-grain bread, whole-grain cereals, and brown rice, have been associated with a reduced risk of developing diabetes, a recent systematic review and meta-analysis of 16 cohort studies found a summary relative risk of 0.68 for three daily servings of whole grains. In the study, it was further discovered that fruits and vegetables as well, including root vegetables, green leafy vegetables, blueberries, grapes, and apples have also been linked to lower diabetes rates. Researchers stated that legumes, plants coming from the pea family, have also been shown to "ameliorate insulin resistance and protect against metabolic syndrome", while greater nut consumption has also been linked to lower diabetes risk. In general, though, "cereal fiber appears to be especially productive against type 2 diabetes" (7).

As the main focus of this chapter's study is to raise awareness about diabetes and to assess whether a plant-based diet has an effect on diabetes, it should be reasonable that a possible specific solution is proposed in regard to a plant-based diet. As the null hypothesis has already been rejected, allowing us to conclude that there is a significant difference between a plant-based diet and a non-plant-based diet and its effects on diabetes, a specific recommended plant-based diet will be proposed for those interested in safeguarding their health and preventing diabetes.

A healthy plant-based diet should contain but is not limited to, "2-3 servings (1 cup raw or ½ cup cooked) of leafy, green vegetables, 2-4 servings (1 piece of ½ cup) of fruit, 2-3 servings (½ cup) of legumes, as well whole grains with 6-11 servings (½ cup cooked or 1 slice whole grain bread), eaten on a daily basis" (8). A healthy plant-based diet, as shown by this chapter's research, does indeed have an effect on the rate of Type 2 Diabetes in humans, and as such, is something that is to be seriously considered by readers.

As such, a plant-based diet, through the combination of the specific benefits given by whole grains, fruits, vegetables, legumes, and nuts, as well as lower insulin resistance promoted by the plant-based diet, coupled with the meta-analyses and chi-squared analysis that resulted in the rejection of the null hypothesis, demonstrates the fact that a plant-based diet does indeed decrease the rate of Type 2 Diabetes.

A plant-based diet is not just a dietary choice; it's a lifestyle that can profoundly impact day-to-day life in numerous positive ways. While it's well-established that a plant-based diet reduces the risk of diabetes, its benefits extend far beyond this particular health concern.

Beyond diabetes prevention, one of the most noticeable and immediate benefits of adopting a plant-based diet is increased energy levels. Plant-based foods are rich in complex carbohydrates, vitamins, minerals, and antioxidants, all of which provide sustained energy throughout the day. Unlike the quick energy spikes and crashes associated with processed foods, a plant-based diet keeps you feeling constantly energized, enabling you to stay alert and focused during your daily activities.

Moreover, a plant-based diet can significantly improve digestive health. The high fiber content of plant-based foods promotes regular bowel movements, prevents constipation, and maintains a healthy gut microbiome. A well-functioning digestive system means less discomfort and more efficient nutrient absorption, ultimately enhancing overall well-being.

The benefits of a plant-based diet extend to cardiovascular health as well. Studies have consistently shown that individuals who follow a plant-based diet have lower risks of heart disease, reduced cholesterol levels, and healthier blood pressure. These factors directly translate into a lower likelihood of heart-related issues in day-to-day life, such as chest pain, fatigue, and shortness of breath.

Maintaining a healthy weight is another advantage of a plant-based diet. With its emphasis on whole, nutrient-dense foods, this diet naturally encourages weight management and may help shed excess pounds. Achieving and maintaining a healthy weight not only boosts self-esteem but also reduces the risk of obesity-related complications like joint pain and sleep apnea.

The anti-inflammatory properties of plant-based foods can also lead to a reduction in day-to-day discomfort. Chronic inflammation is linked to various health conditions, including arthritis and autoimmune disorders. By consuming an abundance of anti-inflammatory foods, individuals on a plant-based diet often experience less joint pain and inflammation, leading to improved mobility and an enhanced quality of life.

Mental health benefits are another noteworthy aspect of a plant-based diet. Research suggests that this diet may reduce the risk of depression and anxiety. Nutrients like folate, antioxidants, and omega-3 fatty acids found in plant-based foods support brain health and mood regulation. Feeling mentally well and resilient is crucial for coping with the stresses of daily life.

In addition to the health-related benefits, a plant-based diet has a positive impact on the environment. Choosing plant-based foods over animal products can reduce one's carbon footprint, conserve natural resources, and contribute to the sustainability of our planet. This aligns with a growing awareness of the need for eco-conscious choices in day-to-day life to address climate change and protect the environment for future generations.

As such, a plant-based diet offers numerous benefits that extend beyond diabetes prevention. From increased energy and better digestive health to improved cardiovascular well-being and mental health, the advantages of this dietary lifestyle are evident in day-to-day life. Additionally, the positive environmental impact underscores the broader implications of choosing a plant-based diet for personal health and the well-being of the planet. Consideration of these benefits makes adopting a plant-based diet a compelling choice for those seeking a healthier, more sustainable way of living.

Due to the data gathered from various articles that analyze the effect of a plant-based diet on Type 2 Diabetes by distinguishing between plant-based diet participants and non-plant-based diet participants, a calculated increase in diabetes rate in non-plant-based diet group from the plant-based diet group (~95% increase), the chi-squared analysis that yielded a p-value less than 0.05, signifying the rejection of the null hypothesis (There is no statistical difference between the rates of diabetes between the group eating a plant-based diet and the group eating non-plant-based diet), as well as several supplementary information regarding the specifics of a plant-based diet, it can be concluded that a plant-based diet does indeed have an effect on Type 2 Diabetes, specifically the fact that it lowers the risk of diabetes in those that engage in a plant-based-diet compared to those that engage in non-plant-based diets.

The weakness of this chapter's research project includes the limited amount of original research study data that was utilized for this chapter's meta-analysis. A more accurate and conclusive result would be yielded from more data in additional research studies about the effect of a plant-based diet on Type 2 Diabetes. Limitations also include the fact that the data gathered mainly focused on Taiwanese Buddhists and Northern Americans, rather than global data from all continents. Access to more globalized data regarding the effect of a plant-based diet on Type 2 Diabetes would yield a more accurate result.

The strength of this chapter's research project includes the fact that the researchers controlled for many factors, for instance, sex, ethnicity, education, income, illness, diet, physical activity, demographics, height, weight, BMI, smoking, cancer as well as alcohol, as to ensure the internal validity of the research.

For future investigations, in order to reach a more accurate result and conclusion, more scientific articles will be examined with data gathered from a variety of populations across the world and control for many factors (sex, ethnicity, education, income, illness, physical activity, height, weight, BMI, smoking, etc...) as to ensure both the validity of this chapter's research project as well as to reach a more precise and accurate result.

Currently, the topic of veganism and diabetes is being researched by the scientific community. For example, there has been a growing body of research on the relationship between veganism and diabetes management, especially focusing on Type 2 Diabetes. Recent research has shed light on potential benefits and insights in practical applications, as well as exploring multiple sides of the impact of a vegan diet on diabetic individuals.

A main focus of scientific discussion revolves around the impact of a vegan diet on glycemic control. Many studies have researched on whether a vegan diet can lead to immense improvements in blood sugar control. These research papers are focused on key outcomes like fasting blood glucose, HbA1c levels, and insulin sensitivity. The findings of the research show promise that vegan diets can be a good dietary alternative to help individuals manage their diabetes through controlling blood sugar levels.

Another pivotal feature of diabetes care is weight management, especially for Type 2 Diabetes patients. Numerous studies have explored the role of vegan diets in weight control and whether the weight loss from these diets can be advantageous for individuals with diabetes. Losing excess weight is seen as a way to increase insulin sensitivity and improve glycemic control, forcing it to be a central focus in vegan diets and diabetes.

Furthermore, individuals with diabetes are more likely to have cardiovascular complications, making it necessary to create a comprehensive approach to managing their health. Researchers have therefore studied the impact of vegan diets on multiple cardiovascular risk factors, measuring changes in blood pressure, markers of inflammation, and lipid profiles for individuals adhering to vegan diets. The purpose of the research is to determine whether a vegan diet can both manage blood sugar levels as well as contribute to heart health in those with diabetes.

New research has demonstrated the importance of the gut microbiome in metabolic health, encouraging investigations in how vegan diets change the composition and diversity of the gut microbiota. Researchers look to unveil possible connections between the gut microbiome and managing diabetes, providing worthwhile knowledge into the effects of vegan diets.

As diabetes is a chronic disease, multiple studies have been interested in the long-term outcomes of vegan diets on diabetes outcomes. Besides short-term glycemic control, research has managed the potential of vegan diets to mitigate complications associated with diabetes, like neuropathy, retinopathy, and nephropathy, through multiple years.

In addition, understanding the behaviors in adopting and maintaining a vegan diet is important. Scientists have looked into factors like patterns of vegan eating, dietary adherence, and the role of social support in successfully implementing and keeping a vegan diet. This knowledge is invaluable for healthcare professionals and dietitians when teaching individuals with diabetes on incorporating veganism as a part of their lifestyles.

Finally, other research papers focus on comparing different diets, focusing on assessing how vegan diets compare with other diet types for managing diabetes. These studies focus on comparing vegan diets to alternatives such as low-carbohydrate diets or the Mediterranean diet, aiming to find out which kind of diet is more helpful for certain individuals with diabetes. Comparative research like this helps dietary recommendations in individual preferences and requirements.

Now, research in the connection between veganism and diabetes is a multifaceted field, with insights into possible benefits and practical considerations of adopting a vegan or plant-based diet for diabetic patients. This research highlights the significance of personalized dietary approaches in managing diabetes, taking into consideration aspects like glycemic control, weight management, cardiovascular health, microbiome composition, long-term outcomes, behavioral factors, and comparative effectiveness.

CHAPTER 4

VEGANISM AND CANCER

Cancer: we've all heard of the disease before. Many forms of cancer exist. It is very dangerous if left untreated and can happen to anyone at any moment. The World Health Organization estimates that around 19.3 million cases of cancer happen each year and around 10 million cancer-related deaths. It is the second highest cause of death behind heart disease. But what exactly is this dangerous disease?

Cancer is the uncontrolled growth of cells, and those cells spread to different parts of the body. The cells go through cell division and duplicate their DNA. DNA is the cell's genetic information that gives the cell step-by-step instructions on how to make essentials, reproduce, and survive. These cancer cells have DNA that has faults in them that impair certain functions or change them. Normally, the cells have certain checkpoints where they can check for these errors, however, these cancer cells don't have these certain checkpoints. They are left to continue to multiply and spread throughout the body.

In the human body, we have many important proteins that give us the ability to walk, think, digest, and all around, exist. How do we get these proteins in our body? The DNA that our cells contain has instructions on how to make these proteins. With these mutations that occur in the cancer cells, the cells won't have the ability to make these proteins and it harms the body. As the healthy cells die off and get replaced with the cancer cells, the body will experience more and more of a lack of these vital proteins, and eventually, death will occur.

There are more than 200 types of cancers that exist. They are classified by the origins of where the cancer cells begin. The main types of cancers that will be discussed throughout the chapter are going to be skin cancer, prostate cancer, and breast cancer. These three kinds of cancers are the three most common throughout the world. [1]

Part 1: Introduction to Various Cancers and Diets to Prevent Them

Introduction to Skin Cancer

Our skin serves as a robust against daily environmental challenges such as injury, disease, and UV rays. In this section, we will focus specifically on skin cancer among others, as a condition that affects millions worldwide; we will explore potential connections between a vegan diet and skin cancer prevention.

The most common preventative measures taken to combat skin cancer involve avoiding intense sun exposure and using protective sunscreen and clothing. Visiting tanning salons is also a no-go as they expose you to UV rays, similar to the effects of sunlight. However, in addition to these strategies, a vegan diet and healthier eating are beginning to show promising results.

A plant-based diet with a lot of antioxidants has been proven to be useful in fighting away various cancers including skin cancer. This diet emphasizes the supply of micronutrients, all with their independent benefits. Vitamin C, a vitamin abundant in citrus fruits, strawberries, and tomatoes, is important for the immune system and collagen synthesis, the formation of connective tissue, which may influence tissue growth. Vitamin E, which is found in wheat germ and almonds, hinders lipid peroxidation, and that can potentially help reduce cancer rates.

Vitamin A (which is found in leafy greens, fish oil, and dairy products) is involved in skin cell proliferation, differentiation, and maintenance thus preserving the skin structure. Along with those highlighted micronutrients, flavonoids, carotenoids, and other assorted substances are also capable of lowering risk for skin cancer. [2]

Chemical reactions are responsible for why these foods are linked to skin cancer prevention. Highly reactive free radicals that lack a pair of electrons can damage cells by producing chemical reactions leading to mutations and alteration of critical macromolecules like DNA, lipids, and proteins - with cancer possibly resulting. In comparison with normal cells, the majority of tumors will display high levels of free radicals.

To wipe out the negative consequences of free radicals, cells have synthesized antioxidants like glutathione. Vitamins such as ascorbic acid and tocopherol offer supplementary reinforcements of the defense array of cells, reducing the risk of skin cancers and various other diseases.[2]

Skin Cancer

Skin cancer is the most common cancer in the US. It is the out-of-control growth of abnormal cells in the epidermis, the outermost skin layer. It is caused by unrepaired DNA damage that triggers mutations. These mutations change the cell's genetic information and harm the cell's ability to function throughout the body. Despite its widespread occurrence, affecting nearly one in five Americans, skin cancer remains one of the most preventable cancers.

What causes these mutations that cause skin cancer? The main culprit is ultraviolet light. Ultraviolet light is a type of electromagnetic radiation that is harmful if exposed in significant amounts. This light makes the cells more prone to mutations and even DNA strand breaking. This leads to a loss of genetic material, and if the information lost were important instructions, then the result will be catastrophic.

Ultraviolet light exposure happens daily. The light and radiation that comes from the sun are in the form of ultraviolet light. That is why we use sunscreen or sunblock to help minimize the damage from sunburns caused by ultraviolet light. Because of these reasons, it is recommended to avoid spending long periods outside in the day. Another way that people get exposed to UV light is through the use of tanning beds. These tanning beds expose you to UV rays, similar to those from the sun, leading to skin cancer. It is six times more likely that someone would develop skin cancer if they visited tanning salons ten or more times before the age of 30. Although it is not guaranteed that not using tanning beds and always wearing sunscreen prevents the development of skin cancer, it is the best way to lower your chances of it happening.

Another important thing is to limit the amount of alcohol intake. Studies have found that excessive alcohol drinking leads to a higher rate of sunburn, and thus a higher chance of developing skin cancer. The breakdown of alcohol in our body generates a lot of free radicals that use up all the antioxidants in our body that protect our skin from the sun. Another reason is that people who drink heavily tend to ignore their health and may not notice the early symptoms that should be detected. [3,4]

Four Main Skin Cancers

The first one is basal cell carcinoma. This is an abnormal and uncontrolled growth of the basal cells in the epidermis. This is often found in areas most directly exposed to the sun. It is mostly caused by long-term exposure to UV radiation and occasionally metastasizes (spreads to the rest of the body). This is the most common type of skin cancer and is rarely fatal. The second kind of skin cancer is squamous cell carcinoma. This is an uncontrolled growth of abnormal squamous cells in the epidermis. Like basal cell carcinoma, it is caused by long-term exposure to harmful UV radiation from the sun and tanning beds. It is commonly found in areas exposed to the sun. This is the second most common form of skin cancer. The danger of it is that it grows rapidly and metastasizes if not detected and treated early. The third kind of skin cancer is melanoma skin cancer. This is a cancer that develops from melanocytes, which are the skin cells that give your skin its color. This form of cancer can appear anywhere in your body. This is the third most common kind of skin cancer and is the most dangerous of them all. Around half of the cases are invasive while the other half are noninvasive. melanoma skin cancer often resembles a sort of mole and is curable if early detection and treatment are done.

The fourth kind of skin cancer is merkel cell carcinoma. This is another dangerous kind of skin cancer that comes from a virus called the merkel cell polyomavirus. Some tumors usually appear as firm, painless lesions or nodules on a sun-exposed area. This form of skin cancer is rare but extremely aggressive. If not detected and treated early, there is a high risk of it metastasizing. [3]

Case Study

The first case study that will be discussed is a 50-year-old Saudi patient. He had complained about swelling in his left foot near the heads of the metatarsals. It had been there for the last two years and the swelling had only gotten worse and was increasing in size. His previous conditions are that he is on diabetic control and treatment. There were no enlarged lymph nodes at popliteal areas or groin, chest X-rays were normal, and his left foot had no bony abnormality

The tumor they had found was made up of epithelioid and spindle cells arranged in diffuse sheets and nests. There were also intracellular melanin pigments discovered in large areas. The cells were large with pleomorphic, hyperchromatic nuclei. This means that the cancer cells were growing in different sizes and that the nuclei were experiencing discoloration. His condition was put in the high-risk category.

A second surgery was carried out and it was discovered that the defect was covered by a split thick skin graft. That is, they found that the cancer cells were a part of the epidermis and also a portion of the dermis. Luckily, the issues were eventually resolved and he recovered.

This goes to show how important it is to get early detection and treatment. Had he not gone to the hospital about his issues, the disease would've gotten more serious and eventually unrecoverable. Malignant melanoma is rare in Saudi Arabia, although the patient likely experienced intense sun and UV light that may have caused the skin cancer. [5]

Vegan Diet

Candidly speaking, to achieve the best chance of reducing the odds of developing skin cancer, one must not only avoid tanning beds and UV exposure, but also consider alternative dietary options. There has been evidence that shows that a plant-based diet that includes lots of antioxidants can reduce the risk of various cancers, including skin cancer. A plant-based diet rich in fruits and vegetables and low in fat has shown promising results in altering skin cancer development.

The issue for most Americans isn't that they don't have access to these resources. As a matter of fact, only around 14.8% - 19.1% of skin cancer survivors are meeting the American Cancer Society's recommendation of at least five servings of fruits and vegetables each day. Only a small fraction of people with skin cancer are actually meeting the suggestions. Most people don't know that fruits and vegetables provide the best polypharmacy against the development of cancer.

Many of these micronutrients that are commonly found provide many outstanding benefits including:

- Vitamin C, which is found in citrus fruits, strawberries, and tomatoes, boosts the immune system and helps with protein synthesis in connective tissue
- Vitamin E, which is found in wheat germ, sunflower seeds, and almonds, helps to prevent lipid peroxidation, which has been known to cause cell membrane damage, and can also potentially reduce cancer rates.
- Vitamin A, which is found in leafy greens, fish oil, and dairy products, plays a crucial role in skin cell proliferation, differentiation, and maintenance. This vitamin can also increase the rate at which the skin heals itself.

Many of the mentioned micronutrients have also demonstrated potential in decreasing skin cancer risk. The reason behind this superpower is in the chemistry. Lacking these micronutrients allows for the release of free radicals, which are molecules with unpaired electrons in their orbitals. These can damage DNA and other critical macromolecules like lipids and proteins. At high enough levels, these free radicals can lead to cancer.

To counteract the harmful effects of free radicals, cells use synthesized antioxidants such as glutathione. Antioxidants like Vitamin C and Vitamin E also help to contribute to and fortify the cell's defense mechanisms, greatly mitigating the risk of skin cancer and other diseases. All of these benefits of prevention and reduction of skin cancer come from having a simple vegan diet that includes the vitamins and nutrients mentioned prior. [6]

Case Study

The next case study is about a patient who survived melanoma skin cancer while being on a plant-based diet. Her diet mainly consisted of foods and ingredients that are known for being cancer-risk-lowering. She claimed that it doesn't hurt her to intake a lot of foods with naturally high levels of fiber, antioxidants, and phytonutrients. She also claimed that she didn't need to consume any dairy for calcium and meat for protein and iron and that she was better off without them. Lastly, she declared that her transition into a plant-based diet was the most joyful and meaningful decision that she had ever made. [7]

Breast Cancer

What is breast cancer? Breast cancer is one of the main forms of cancer. Its origins begin in the breast and can start in one or both breasts. It may seem that breast cancer only occurs in women, but in reality, it can happen to men. Breast cancer can start in any part of the breast tissues.

Breast cancer spreads when the cancer cells get inside of the blood or lymph system and diffuse throughout the body. If the cancer cells have reached your lymph nodes, then there is a chance that the cancer has metastasized to the other parts of the body. Not all women with cancer cells in the lymph nodes develop metastases, as metastases can develop later on. Symptoms of breast cancer include:

- Skin dimpling
- Breast or nipple pain
- Nipple retraction
- Nipple discharge other than breast milk

- Swelling of the breast
- Swollen lymph nodes under the arm or near the collarbone
- Nipple or breast skin that feels dry and is thickened and flaking

One important thing to note is that any sort of lump or mass on the breast that had not been previously there should be checked out. It's not guaranteed that you will be able to see and know when exactly you have breast cancer, so routine screening is important just in case. [8,9]

Melittin & Breast Cancer

The world has many different treatments for cancer. It has been a hot topic lately in the race to eventually find a cure for cancer. One thing that had been discovered recently was the usage of melittin (bee venom), to try and suppress cancer cells. It was shocking to the world that the same compound that gives you pain from bee stings also may have a shot at curing cancer. The prospect of melittin eventually being used regularly is still in its early stages, as more testing needs to be done. This portion of the chapter will go over details on how melittin works and some trials that have been done.

It is not just now that medical practitioners have been using bee venom for pharmaceutical uses. The usage of melittin dates back at least 5,000 years to the ancient Egyptians, Chinese, and Greeks. They would use this to treat many things like joint inflammation and pain. Another thing is that the honey from bees has antibacterial properties that help with treating wounds, easing indigestion, and embalming the dead.

The venom that the honeybees inject when they sting is a complex mixture of different proteins, enzymes, sugars, lipids, and other substances. The most abundant of these is melittin, which is responsible for a lot of the medicinal properties of bee venom. [10]

So how does this work? The peptide melittin has already been discovered to have properties that allow it to break down lipid membranes and disrupt the growth of cancer cells. It does this by disrupting the signals cancer cells use to replicate itself. This in turn slows the growth of cancer cells.

One study showed that at a certain concentration of melittin, the serum had killed all the cancer cells in their test within an hour. The difference between this form of treatment and others is that regular healthy cells sustained minimal damage. Researchers also have found that melittin is extremely effective at destroying certain types of cancer cells, including some subtypes of breast cancer.

How does melittin kill cancer cells? What melittin does is that it suppresses cell proliferation and also induces apoptosis, which is programmed cell death. Apoptosis plays an important role in removing damaged cells. Melittin caused cancer cell apoptosis by adjusting the genes related to apoptosis such as p53, Bax, and Bcl-2. By doing this, apoptosis is induced and the cancer cells start to die off.

One example of melittin in action is when bee venom was used on MDA-MB-231 breast cancer cells. The results showed that the cancer cells began to die in a process that denatured and degraded the proteins the cancer cells had. It also fragmented the cancer cell's DNA, making it unable to reproduce.

Other things about melittin are that it is amphipathic and capable of disrupting the integrity of the tumor cell membrane bilayers which lead to holes and flaws in the bilayer. [11,12]

Melittin works on most cancers, however, studies have shown that breast cancer cells were more effectively eliminated in the experimental group than in other cancer cells in the control group.

Cancer cells may eventually develop resistance to this form of treatment. For even better results, melittin pairs well with other forms of treatment. For example. Doxorubicin and melittin given together have proven to have synergistic effects on breast cancer cell proliferation. One thing to note is that melittin does have some side effects. It is known to cause nonspecific cytotoxicity (cell poisoning) and hemolysis (the destruction of red blood cells). So, it's important to not overuse melittin. [11]

Two Breast Cancers Melittin Works Well For

The two breast cancers that melittin works well for are HER2-Positive Breast Cancer and Triple-Negative Breast Cancer.

HER2-positive breast cancer is characterized by an overexpression of the human epidermal growth factor receptor 2 (HER2) proteins. These are the proteins that control cell division and growth, leading to uncontrolled growth. This leads to tumors forming.

HER2-positive breast cancer tends to grow faster, metastasize, and recur, but also it responds to treatments better. This form of cancer does not have any symptoms, although it can be detected through screening. Diagnostic tests, like immunohistochemistry (IHC) and fluorescent in situ hybridization (FISH), are also good methods for diagnosis and status reports.

These tests measure the amount of HER2 protein levels and also search for any extra copies of the HER2 gene.

Hormone receptor testing is also performed to test if hormone therapy is an option. If the tests show positive results for hormone therapy and tumors, then the patient will benefit from both targeted therapy and hormone therapy. These treatments aim to try and slow or halt the tumor growth and reduce the chance of spread and recurrence.

Early detection of HER2-positive breast cancer is vital as this form of breast cancer is usually more aggressive than other forms. Early detection also allows for the appropriate treatment to be done for the patient. The survival rates are generally high for the patients, at around 90%. [13]

Triple-negative breast Cancer accounts for around 10-15% of all breast cancer cases. This form of breast cancer is sort of the opposite form of the HER2-positive breast cancer, the HER2 protein is barely expressed. These cancer cells also don't have estrogen and progesterone receptors. This makes treatments like hormone therapy or targeted HER2 drugs not work. [14]

The danger of this cancer is that it grows and spreads faster and has less treatment than others, making the prognosis worse. The five-year survival rates for different groups of triple-negative breast cancer are as follows:

- Localized (unspread) - 91%
- Regional (spread to nearby structures and lymph nodes) - 66%
- Distant (spread to different body parts) - 12%
- Overall - 77%

For both these breast cancers, melittin is effective in destroying and eliminating these cancer cells. Melittin is relatively easy to obtain and not very expensive, so it may be an extremely viable option for cancer treatment in the future.

Case Study

Practice research into the usage of melittin on humans is still in the early stages of development. However, melittin has been studied in mice and the results are promising. From the results, melittin killed cancer cells in culture dishes as well as in mice. The research also found that cells from the HER2-Positive Breast

Cancer and Triple-Negative Breast Cancer were killed more effectively than others.

Melittin was also modified and combined with other drugs to enhance the effectiveness of killing the cancer cells. In the study, the chemotherapy drug docetaxel was combined with melittin and was tried in the mice. The results showed that the combination had killed more breast cancer cells than docetaxel alone. [15]

All of these findings and research done on how melittin kills cancer cells support the idea that bee venom could be an effective method for treating and even curing cancer in the future. The current situation is that there must be more research done and experiments with humans to test the effectiveness. The situation is promising for the future.

Vegan Diet

Having a vegan diet composed of fruits, vegetables, grains, and beans promotes breast health. Foods containing high amounts of fiber, minerals, and vitamins lower breast cancer risk. Many plant-based foods also have high amounts of Vitamin C, beta-carotene, and selenium, all of which are proven to lower cancer risk. These resources also help to eliminate excess estrogen which lowers the chance of cancer from developing.

Dairy products have a high fat content. These are linked to breast cancer. A study was conducted and found that out of a 10,000 women sample, the women who consumed a low-fat diet had a 23% lower chance of getting breast cancer again and a 17% lower risk of dying. Another study after comparing the diets of women with and without breast cancer discovered that those who had consumed dairy products like cheese had a 53% higher risk for breast cancer.

Also, the World Cancer Research Fund found that women who consumed between ¼ and ⅓ cups of cow's milk a day had a 30%

higher risk of getting breast cancer. One cup of milk a day had a 50% higher chance of getting breast cancer, and 2-to 3 cups a day had an 80% chance of getting it.

Red and processed meat have also been shown to have a strong correlation with breast cancer. According to The Nurses' Health Study II, it was found that premenopausal women who consumed red meat had a 97% higher risk of a common type of breast cancer. Another study indicated that women who ate mostly red meat had a 25% higher chance of breast cancer than women who rarely ate red meat. [8]

All in all, for women looking to minimize the risk of obtaining breast cancer, the suggestion is to have a plant-based diet with a minimal amount of red meat and dairy intake. Getting breast cancer is not fully preventable, but having a vegan diet is the best way to reduce the chances.

Prostate Cancer

Prostate cancer is cancer of the prostate. It begins when the cells found in the prostate begin to grow out of control. Most prostate cancers are adenocarcinomas. These are cancers that develop from the gland cells that make prostate fluid. Some prostate cancers grow and spread quickly while others are slow. Some are aggressive and some require little to no treatment.

Prostate cancer is the most common cancer for males other than skin cancer and is one of the leading causes of death in males. It is very important to notice the symptoms and receive treatment as soon as possible. Some symptoms include

- Trouble urinating
- Decreased force in the stream of urine
- Blood in the urine

- Blood in the semen
- Bone pain
- Erectile dysfunction
- Weight loss without trying

One thing about prostate cancer is that the cause is unknown, although many factors can affect prostate cancer in general. Genetics, the place where you were born, radiation, and smoking are all factors that may play a role in the development of prostate cancer. However, many of these are out of one's control. One thing that significantly affects prostate cancer rates is one's diet. [16]

Part 2: The Mediterranean Diet

Besides the diets mentioned above, the Mediterranean Diet will be described in-depth and possible connections to cancer will be investigated.

The Mediterranean Diet (MD) has been shaped by the culinary traditions of the Mediterranean regions and is currently considered as a health leader. It has been documented to be able to cut the chance of cancer, and therefore, the available research should be reviewed.

A collection of 13 randomized controlled trials was found that reported the effect of strict adherence to the MD on cancer mortality. The studies were carried out using different scoring systems for MD adherence, such as the Mediterranean Diet Score (MDS), alternate MD (aMED), modified MD (mMSD), MD pattern (MPD) and relative Mediterranean diet (rMED) scores. They were all cohort studies; they tracked groups of subjects over the period of time in order to investigate the link between diet and cancer [17].

To ensure that their results were reliable, these studies used dietary questionnaires that have been tested for accuracy and depended on trusted sources like mortality and cancer registries. They also examined other elements that could affect the results, including age, gender, area, calorie intake, physical activity, education, and BMI.

In regards to MDS-related adherence and cancer mortality, some studies reported a considerable favorable impact, while others failed to demonstrate such effect. For example, the European Prospective Investigation into Cancer and Nutrition (EPIC) documented the multi-country investigation in which altogether 450,000 people participated. The investigation revealed that both Mediterranean and prudent dietary patterns (MDS and rMED, respectively) could be related to a reduced risk of mortality from any cancer.

On the other hand, a Switzerland study with over 17,000 men and women showed that higher MDS adherence reduced cancer mortality in males but not in the case of females. Some other studies did not even confirm a direct association between mortality rates due to cancer and MDS.

When other MD indices like aMED, mMED, and rMED were explored, there were some marked differences. Across nine studies under the aMED regime, there was always a spark of hope. Compliance with aMED was associated with decreased mortality rates caused by cancer.

In a study with ~38,000 Swedish women who follow mMED, there was a lower risk of cancer mortality. One the studies of rMED had opposite results though. A study among the EPIC cohort has found out that the risk of cancer mortality has been lowered. Another study in Spain has seen no notable impact on the overall cancer mortality rates.[17]

In summary, when specifically investigating the MD, the data is sort of a mixed bag. Some indices like aMED show that it is in a way reducing the chances of cancer mortality but there is no conclusive answer yet.

In the pioneering study completed by Tantamango-Bartley et al. in 2010 [5], the researchers investigated the relationship that various diets (non-vegetarian, lacto-, pescetarians, vegans and semi-vegetarians) had with the incidence of cancer.

The study, known as AHS-2, had a sample size of ~70,000, and was conducted to explore the diet and lifestyle contributions to cancer occurrence. Respondents took a questionnaire that embraced hundreds of foods, including frequency, serving sizes and dietary habits for the last year to establish the dietary status. Along with diet information, the questionnaire gathered other important metrics: demographics, medical and family history, exercise, smoking, and alcohol.

During the 4.14 year average follow-up period, a total of 2939 cancer cases were identified in both males and females. There was evidence of a statistically significant decreased risk of overall cancer among vegetarians. And although not as significant, the effect was notable in males. The vegetarians, especially lacto-vegetarians, experienced the most reduction in the risk of developing colon and rectum cancer. In line with this study, vegans are getting significant protection against female specific cancer as well.[18]

Part 3: Plant-Based Diet Effects on Prostate Cancer - Author Research

The research that is going to be discussed is about a plant-based diet and its effects on prostate cancer. It is a meta-analysis, meaning it combines data from previous research and produces a paper based on the results.

Average Rate for Prostate Cancer

For a non-plant-based diet, the articles that were found to be useful contain information and data from all ages. An article by ASN found that 1079 patients got it out of a total of 26,346. Seventy-eight cases came from people less than or equal to the age of 50, 213 between the ages of 50 and 59, 298 between the ages of 60 and 69, and 390 above the age of 69. It was also found that out of 214,974 people in the study, 9437 got prostate cancer. This means the rate should be around four people per 100 or 4% of the population. [19]

For a plant-based diet, one article stated that out of 2140 patients in the study, 74 got prostate cancer. One more extensive study found that out of a population of 27,188, 333 got prostate cancer. The rates were around 2%, so anything too much or too little was excluded from being used as data. [19]

After obtaining the useful information, the total number of human subjects for the 2 groups, one with a regular diet and one with a vegetarian diet, was noted. Also, the number of prostate cancer cases per group was noted. Then, calculations were made to find out the rate at which the human subjects were getting prostate cancer for both groups.

Then, it must be determined whether the difference in the rates between the 2 groups is statistically significant. Through a chi-squared calculation, it is determined that the p-value is less than 0.05, and therefore the difference in the rates between the 2 groups is statistically significant.

Results

After having a significant enough data set, the sum of the total number of people for each group was added up. Then, the number of cases and non cases was combined, and the values from this group were divided over the total number of people to get the rough percentage of people that got prostate cancer for each group. Getting this percentage was crucial in calculating how many people would get prostate cancer in each group.

In a chi-squared calculation, expected and actual values are required. The actual value would come from the number of people who got prostate cancer from the plant-based diet group, and the expected value would come from the number of people who got prostate cancer from the non-plant-based diet group. Then, one should subtract the actual value from the expected value and square the difference between them. Lastly, the result was divided by the expected value to get a specific number.

Table 1

Expected Value	Individuals	Cases	Noncases	%	Cases Per 100,000
	214974	9437	205537	4.389833	
	190543	7492	183051	3.931921	
	26346	1079	25267	4.095498	
	47781	1369	46412	2.865156	
Total	479644	19377	415224	4.039871	4039.871238
Actual Value	2963	64	2899	2.159973	
	2140	74	2066	3.457944	
	5486	100	4524	1.822822	
	27188	333	26855	1.224805	
Total	37777	571	36344	1.511502	1511.502

In Table 1, the participants in the non-plant-based diet group were summed up, which had 479,644 subjects. Then, the number of people that got prostate cancer was summed up, which was 19,377. Dividing that number by the total number of subjects would get the approximate chance that a non-plant-based diet would cause prostate cancer, around 4.04%. After doing the same with the plant-based diet group; out of the 37,777 subjects, 571 got prostate cancer. The chance of getting prostate cancer was around 1.51%.

The chi-squared calculation requires numbers and not percentages, so the rates would have to be used to find out how many people, on average, would get it in each group with equal amounts of population. Then, making both populations 100,000 allows the calculations of the number of people that get prostate cancer to be easier. Using the percentage from the expected group, 4.04%, the percentage was first divided by 100 to turn it into a decimal. That result is 0.0404. That shows that 0.0404 would usually get prostate cancer out of a one-person population. To change the population to 100,000, multiply 0.0404 by 100,000 to get 4040 people. The expected value is 4040 people would get prostate cancer with a non plant-based diet.

The exact process was used to calculate the actual value. The percentage for people with the plant-based diet was 1.51%, so dividing that by 100 to get the decimal would be 0.0151.

Then, multiplying it by 100,000 to change the population, and about 1510 people would get prostate cancer out of 100,000 people with a plant-based diet.

Table 2	Expected Cases	Actual Cases	Actual-Expected	Squared	Divided by Expected
	4040	1510	-2530	6400900	1584.38118

Conclusion

The last step of the chi-squared calculation is to plug the values into the equation. The equation for the chi-squared calculation is ((actual value - expected value)^2)/expected value. Putting the numbers in, the result would turn out to be ((1510-4040)^2)/4040), which is around 1584.4. Using the number that was obtained from the formula, it can be used to find the p-value. Putting it into a calculator, the p-value is less than 0.05. Using this result, it can be concluded that the data present has shown a significant effect.

In conclusion, there has been strong evidence suggesting, and as tested, the data from several articles proved that prostate cancer rates are significantly lower compared to a regular, non plant-based diet.

To summarize, cancer is no easy disease to defeat. It's a pain to deal with and a pain to have to worry about. From the research mentioned throughout this chapter, there is a clear-cut way to minimize the chances of getting cancer or reducing its effects: that is, picking up a plant-based vegan diet. There are also many new developments and treatments for cancer such as melittin. Although it needs more research, it brings the world closer to finally discovering the cure to the disease that plagues millions of people.

CHAPTER 5

ALZHEIMER'S DISEASE

Part 1: What is Alzheimer's Disease?

Alzheimer's disease is a devastating and irreversible brain disorder that affects memory, thinking, behavior, and language that is a global health concern, affecting millions of people worldwide [1]. It's characterized by a gradual deterioration of speech, memory, and language and is the leading cause of dementia, which represents about 70% of all cases [2]. And according to the World Health Organization (WHO), an estimated 50 million individuals were living with dementia, including Alzheimer's, in 2020, with nearly 10 million new cases reported every year. In the US alone, an estimated 5.8 million individuals were living with Alzheimer's disease as of 2020 [2].

Alzheimer's disease is idiopathic, meaning that the exact cause of Alzheimer's is unknown, but it is believed to be due to multiple factors, which would take into account the genetic, environmental, and lifestyle factors that possibly contribute to the development of Alzheimer's [3]. The disease is generally characterized by an accumulation of two abnormal proteins in the brain: beta-amyloid plaques and tau tangles [4]. Beta-amyloid is produced naturally in the body and is usually broken down and eliminated by the brain [4]. However, in those with Alzheimer's disease, beta-amyloid peptides accumulate and form abnormal plaques outside and around neurons in the brain [4]. This accumulation of beta-amyloid plaques in the brain can trigger an immune response, releasing inflammatory molecules that can damage and destroy brain cells, on top of interfering with the communication between neurons, causing neurons to eventually die [5]. In addition, the formation of tau tangles, another characteristic of Alzheimer's, can be triggered by this accumulation of beta-amyloids [6]. Tau tangles are essentially abnormal clumps of a protein called tau that can accumulate inside neurons in the brain. Tau itself is a protein that helps to stabilize the structure of neurons and transportation of nutrients/molecules within the cell [6]. When tau begins to clump together, they are typically found around areas in that brain that are important for memory and cognitive function and can trigger genetic mutations, chronic inflammation, and oxidative stress [5].

Alzheimer's disease can be characterized by several stages, with the symptoms worsening quickly over time. Early stage symptoms include mild forgetfulness, difficulty in recalling recent events, and problems with language and communication. And as the disease progresses, the symptoms may become even more extreme, including severe memory loss, confusion, disorientation, mood swings, and difficulty with everyday activities such as dressing, bathing, and eating [7].

Alzheimer's disease, in general, affects both men and women, but it has been observed that women are disproportionately more affected. Almost 2/3 of individuals with Alzheimer's are women. A potential reason for this statistical difference between genders may be due to women's longer life expectancy, as age does serve as a significant risk factor. Most people who do develop Alzheimer's disease are 65+ years old, however, there is also a phenomenon called "early-onset Alzheimer's," which can affect individuals as young as their 30s or 40s and will be discussed within a case study later on in this chapter [7].

Alzheimer's disease progresses through several stages, with the signs and symptoms worsening as the stages progress. The earliest stage is known as preclinical Alzheimer's disease, where there aren't really any noticeable symptoms. This phase can last for years before concerning symptoms begin to present themselves. The next stage is mild cognitive impairment (MCI), which is characterized by slight memory and cognitive changes. People with MCI may experience occasional memory lapses, such as forgetting recent conversations or events that would typically be remembered. They might struggle with estimating task duration or planning complex activities. However, it's important to note that not everyone with MCI progresses to Alzheimer's disease [7]. As Alzheimer's advances, individuals enter the stage of mild dementia and the signs and symptoms become more apparent, including memory loss of recent events, difficulty with problem-solving, and changes in personality. A few other examples of how the symptoms may present itself includes

- Planning tasks or managing finances becomes overwhelming
- Expressions of irritability or withdrawal
- Organizing thoughts and conveying ideas become challenging
- Spatial navigation issues (ex. : misplacing belongings)

In the moderate dementia stage, confusion and forgetfulness worsen, requiring increased assistance with daily activities. Judgment becomes poor, and confusion about time, location, and identities may increase. Help becomes necessary for activities like dressing, grooming, and using the bathroom. Personality changes can include unfounded suspicions, hallucinations, and aggressive outbursts, in addition to wandering behavior becoming more common. As Alzheimer's reaches severe dementia, communication becomes increasingly difficult and fragmented. Coherent conversations become rare, and there's a dramatic increase in dependence on caretakers. Physical capabilities decline rapidly, with difficulties in walking, sitting, and maintaining head posture without assistance. Muscles can become rigid due to a decrease in movement and control over functions like swallowing and bladder/bowel control is significantly decreased. Throughout these stages, Alzheimer's disease has a significant toll on individuals' cognitive abilities and physical well-being, emphasizing the importance of early detection, intervention, and ongoing support for both patients and their caregivers. [7]

Unfortunately, at the same time as it robs the patient of their memories, Alzheimer's places a substantial burden on both individuals and their families. In the United States, it is estimated that in 2020, unpaid caregivers provided 15.3 billion hours of care to individuals with Alzheimer's and other dementias. The economic cost of Alzheimer's and other dementias in the U.S. exceeded $300 billion annually, including both direct medical costs and the value of unpaid caregiving. As the global population ages, the number of people living with Alzheimer's is expected to rise significantly. By 2050, it is projected that the number of individuals with dementia will reach 152 million, with the majority of cases being Alzheimer's disease. [8]

Alzheimer's prevalence varies by region, with some areas experiencing a higher incidence than others. Factors such as genetics, lifestyle, and healthcare infrastructure contribute to these variations. For example, in Western Europe and North America, the prevalence is relatively high, while in some Asian countries, it is lower. While age is the most significant risk factor for Alzheimer's disease, other factors play a role. Genetics, family history, and certain lifestyle factors, such as diet and physical activity, can influence the risk of developing the disease. The presence of specific genetic variants, such as the APOE ε4 allele, increases the risk of Alzheimer's. [9]

As research into Alzheimer's disease continues, ethical considerations become increasingly important. These considerations encompass issues related to the conduct of clinical trials, patient consent, privacy, and the equitable distribution of resources. Alzheimer's research often involves individuals who may have impaired decision-making capacity due to the disease. Ensuring that participants provide informed consent is critical. Ethical guidelines emphasize the need for advanced directives and surrogate decision-makers to protect the rights and welfare of research participants. With the advent of Big Data and the use of advanced technologies in Alzheimer's research, privacy and data security have become paramount. Researchers must adhere to strict data protection protocols to safeguard sensitive information and ensure that it is not misused. Access to Alzheimer's research, clinical trials, and treatments should be equitable. It's essential to address disparities in healthcare access and research participation among diverse populations to ensure that the benefits of research reach everyone. Ethical discussions revolve around end-of-life care decisions, including the use of advanced directives, palliative care, and euthanasia debates. Ethical considerations also extend to public awareness campaigns and education about Alzheimer's disease. It is essential to provide accurate information to the public while avoiding stigmatization and discrimination against individuals with the disease. [10]

It is essential that there are caregivers that can support individuals with Alzheimer's disease. However, taking care of someone with Alzhimer's can be emotionally and psychologically challenging since caregivers will face stress, anxiety, depression due to the cognitive decline of their loved ones. Also, having to physically assist Alzheimer patients with daily activities can be exhausting, with the addition of having to work for extra hours due to patients needing extra assistance. There should be support groups, and educational programs that help caregivers dealing with stress, anxiety, depression. There should also be a policies aimed at providing more flexibility for caregivers.[11]

Part 2: Diet's Effect on Alzheimer's

The Mediterranean diet has been linked to a lower risk of Alzheimer's and dementia. The diet includes fruits, vegetables, legumes, whole grains, salmon, and olive oil as oil. This diet lowers the risk of hypertension, high cholesterol, and type 2 diabetes [12]. "In 2015, Dr. Martha Clare Morris and colleagues at Rush University Medical Center and the Harvard Chan School of Public Health published two papers introducing the MIND diet" [13] A MIND diet score was created to help people identify meals and nutrients, as well as daily serving amounts. For a decade, a group of elderly people were recruited from more than 40 retirement complexes and senior public housing units in Chicago [13].

The diet consists of foods high in nutrients that are thought to lower oxidative stress and inflammation in the brain. Scientists conducted a study in which they used questionnaire data to compare and contrast the two diets. Those who reported eating healthy on the questionnaire received a higher score than those who ate unhealthy meals such as red meat, high-sugar, and high-fat items. Those with a higher score demonstrated less indications of Alzheimer's disease due to lower levels of amyloid plaques. Amyloid plaques are made up of beta-amyloid, which are protein fragments that clump together from the lipid membrane that surrounds nerve cells. As a result, large quantities of this clump protein together to create plagues that collect between neurons and impede cell function [13].

Prospective studies have also shown that limiting alcohol and carbohydrate consumption can help to reduce the risk of Alzheimer's disease. In 2013, experts from the International Conference on Nutrition and the Brain developed a recommendation to help prevent Alzheimer's disease. This involves restricting saturated and trans fats, as well as dairy, meat, and vitamin E supplements. Additionally, try to reduce your consumption of aluminum and metal-containing items. Aluminum can be present in both cooking equipment and foods. According to a 1953 study, people with Alzheimer's have high quantities of iron in their brains. With that said, do not consume too much zinc, copper, or iron because they are linked to protein amyloid and tau tangles, which are protein clumps [14].

Researchers have also looked to the ketogenic diet as a potential treatment and prevention strategy for Alzheimer's disease. The ketogenic diet is a low-carbohydrate and fat-rich diet traditionally used to treat epilepsy. However, because impaired glucose metabolism and the buildup of amyloid beta plaques is associated with Alzheimer's disease, ketones are being considered as alternate metabolic precursors to glucose in the brain. Thus, it is possible that a ketogenic diet will lead to reduced formation of amyloid plaques and their neurotoxicity. Several studies have shown that the ketogenic diet is a promising solution to slowing Alzheimer's disease, but with some results being difficult to interpret, further research will need to be done on the diet's effectiveness and any possible drawbacks, such as the hassles required to maintain such a diet [15].

Part 3: Depression's Effect on Alzheimer's

Depression's Effect

Research suggests a strong connection between depression and the progression of Alzheimer's disease. Multiple studies have consistently shown that individuals with a history of depression face a significantly heightened risk of developing Alzheimer's and experience a more rapid cognitive decline compared to those without depression [16]. Also, It can be really challenging for doctors to distinguish between symptoms of depression and early stage Alzheimer's. For example, a study published in the Journal of Neurology, Neurosurgery & Psychiatry found that individuals with a history of depression had a 2.5 times higher risk of developing Alzheimer's compared to those without such a history. Moreover, a longitudinal study conducted over 20 years, published in Neurology, revealed that individuals diagnosed with depression during this period had a 70% higher risk of developing Alzheimer's later in life. This concerning association raises questions about the potential for depression not only to increase the risk of Alzheimer's but also to accelerate cognitive decline, making it an urgent area of investigation. [17]

The link between mental health conditions, including depression, and neurodegenerative disorders is substantial. A comprehensive meta-analysis published in the Archives of General Psychiatry reported that people with depression had a 4.68 times higher risk of developing dementia compared to those without depression [18]. This finding underscores depression as a significant risk factor for cognitive decline and the progression of Alzheimer's disease, a risk that persists even decades after receiving a depression diagnosis. This long-term risk was confirmed by a study published in the Journal of the American Geriatrics Society, which followed individuals with depression over a 20-year period and found a persistently elevated risk of dementia development. These statistics underscore the critical importance of addressing depression as a potential precursor to Alzheimer's disease. [19]

Depression frequently coexists with cognitive impairment in Alzheimer's patients, further exacerbating the severity of cognitive decline. A comprehensive study published in the Journal of Alzheimer's Disease reported that up to 50% of Alzheimer's patients experience depressive symptoms. These symptoms, including impaired concentration, memory difficulties, and reduced executive functioning, often overlap with those observed in Alzheimer's disease. This overlap complicates the accurate diagnosis and assessment of both conditions, making it crucial to distinguish between depressive symptoms and cognitive decline attributable to Alzheimer's disease. [20]

Depression's impact on the brain's structure and function interacts with Alzheimer's pathological processes. A study in Neurobiology of Aging demonstrated that chronic stress, a common feature of depression, leads to dysregulation of the hypothalamic-pituitary-adrenal (HPA) axis, resulting in significantly increased production of stress hormones. These hormones can contribute to neuroinflammation, oxidative stress, and the accumulation of beta-amyloid plaques and tau tangles, hallmark pathologies in Alzheimer's disease. This research highlights how depression-induced neurobiological changes can synergize with Alzheimer's disease pathologies to accelerate cognitive decline. [19][20]

Additionally, Alzheimer’s disease can increase symptoms of depression due to cognitive impairments and functional problems. A study published in the Journal of Geriatric Psychiatry and Neurology figured that individuals with Alzheimer’s disease had 2.3 times higher risk of being diagnosed with depression in contrast with individuals without Alzheimer’s disease. Individuals experiencing Alzheimer’s disease deal with dependence on family members, impaired social interactions, feelings of resentment, and hopelessness, which results in the progress of depression. [21]

Alzheimer’s disease results in the decline of the normal functions of neurotransmitters such as serotonin, norepinephrine, and dopamine. These neurotransmitters play an important role in regulating mood and emotions. Research published in the Journal of Alzheimer’s Disease claimed that individuals with Alzheimer’s disease often reveal a significant variation of neurotransmitter levels which contributes to the depression symptoms. The neurodegenerative process in Alzheimer’s disease may affect various areas in brain regions such as mood regulations, which results in an incline of depression symptoms. This neurobiological aspect allows researchers to understand the interconnection between Alzheimer's and depression. [22]

Alzheimer's disease and depression significantly affects an individual's quality of life. A study published in the Journal of Alzheimer's Disease found that the presence of depression in Alzheimer's patients is associated with a 38% higher level of functional impairment, 56% increased caregiver burden, and a 47% faster decline in cognitive functional abilities. These statistics imply the need for comprehensive care strategies that address Alzhimer's disease and depression. Depression symptoms can intensify social isolation, hinder participation in activities, and reduce overall well-being. [23]

The Link between Depression and Dementia

Current research, including case control and prospective studies has provided important insights that are associated between depression and dementia. The evidence that supports this connection, with multiple hypotheses that explains this intricate relationship.

Depression as a Prodrome of Dementia

One hypothesis suggests that depression may serve early dementia. Clinical observations have indicated that individuals diagnosed with depression may progress to dementia. This alignment with dementia onset hints the possibility that depression is an early symptom. Two mechanisms have been proposed to explain this connection:

Depression as Prodrome of Alzheimer's: Some evidence indicates that depression may precede the onset of Alzheimer's disease. A study published in the *Journal of Alzheimer's Disease* was able to discover that individuals with a history of depression had a 50% higher risk of developing Alzheimer's later in life. This neuronal loss would need to occur early in Alzheimer's progression, which contradicts its typical late-stage presentation. [19]

Depression as a Prodrome of Vascular Dementia: Another theory proposes that certain cases of late-life depression may be attributed to cerebrovascular disease. Although this disease may not initially occur clinically when depression first appears, MRI studies have revealed a higher prevalence of white matter changes linked to cerebrovascular disease in depressed elderly individuals. These changes are associated with cognitive impairments and an increased risk of progressing to vascular dementia. A study published in *JAMA Psychiatry* found that patients that were diagnosed with depression with white matter changes were more likely to transcend to vascular dementia. [19]

Depression Accelerating Dementia's Clinical Manifestations

Alzheimer's and vascular dementia are part of a spectrum that involves various degrees of neuronal loss. A prolonged preclinical phase usually precedes evident dementia, which becomes apparent once a threshold of neuronal loss is reached. A prominent risk factor that may influence the rate of neuronal loss is the concept of brain reserve. [24]

This hypothesis posits that a larger brain reserve can compensate for neuronal loss, explaining why individuals with higher education levels, greater intelligence, or larger brain sizes are at a reduced risk for Alzheimer's. A study published in *Neurology* found that individuals with more years of education had a lower risk of developing Alzheimer's. This concept may apply to depression as a risk factor for dementia, affecting the threshold for dementia. Depression often comes with significant cognitive deficits, and when combined with dementing diseases, could contribute to the onset of dementia. [24]

Depression and the Glucocorticoid Cascade

A compelling hypothesis proposed by Sapolsky and colleagues centers on the idea that stressors prompt the brain to stimulate the adrenal cortex, leading to the secretion of glucocorticoids. These hormones prepare the body to cope with stress by increasing circulating energy substrates. However, receptors in the hippocampus detect elevated glucocorticoid levels and inhibit further secretion, as prolonged secretion can harm the hippocampus. [25]

With age, a loss of glucocorticoid receptors occurs, impairing feedback inhibition and causing further damage to the hippocampus. Much of the evidence for this theory has come from animal studies, but the work that has been done on humans show that elevated levels of cortisol, (the body's primary glucocorticoid) in elderly people over a five year period can be related to memory performance and hippocampal volume [26]. People with higher cortisol levels were found to have poorer memory performance and greater hippocampal atrophy. Though the participants of this study did not have dementia or clinical depression, hypersecretion of cortisol and nonsuppression has been reported more often in both depression and dementia. Furthermore, atrophy is one of the early brain changes that come with Alzheimer's disease and has been observed in some patients with depression. This cascade of events potentially links chronic depression, stress, and glucocorticoid release to hippocampal damage, which, over time, may contribute to the development of dementia. [27]

Specific Symptom Patterns

While the overarching association between depression and dementia is intricate, recent studies have also examined specific symptom patterns. These investigations aim to uncover whether certain depressive symptoms or patterns are more closely linked to the development of dementia.

Motivation Symptoms: Some prospective studies suggest that individuals who later develop Alzheimer's are more likely to exhibit motivation-related depressive symptoms at baseline. These symptoms include a lack of interest, loss of energy, and concentration difficulties. Notably, mood-related symptoms are less commonly observed in this context. [28]

Cognitive Symptoms: Another prospective study followed participants over an average of 3.2 years and found that cognitive symptoms, such as a slowing of thinking, and depressed mood were the only symptoms preceding the onset of Alzheimer's. However, this pattern was more evident in individuals with higher levels of education. This finding highlights the complexity of the relationship between specific depressive symptoms and dementia risk. [29]

Exploring Novel Avenues in Depression-Dementia Research

In the field of depression-dementia research, there are ongoing investigations that continue to uncover intriguing connections and propose hypotheses for exploration. These emerging areas expand one's understanding of the complex dynamic between depression and dementia, offering promising insights into risk factors and intervention strategies.

Inflammation and Immune Response: Recent studies have shifted to the role of inflammation and the immune response associated with depression and dementia. Chronic inflammation, often observed in depression, has been implicated as a potential contributor to neurodegenerative processes. Research has shown that inflammatory markers, such as C-reactive protein and interleukin-6, contain higher levels in both depression and dementia. Understanding the relationship between inflammation, depression, and dementia paves the way for innovation therapeutic approaches targeting immune modulation. [30]

Microbiome and Gut-Brain Axis: The gut microbiome, a diverse system of microorganisms residing in the digestive tract, has potential impact on mental healthj and cognitive function. Emerging research suggests a bidirectional communication between the gut and the brain, known as the gut-brain axis. Alterations in the gut microbiome competitions, known as dysbiosis, have been linked to depression and dementia. Exploring how gut microbiota influence inflammation, neurotransmitter production, and cognitive function could reveal novel pathways for intervention and prevention protocols. [31]

Sleep Disturbances: Insomnia and sleep apnea have been recognized as potential contributors to both depression and dementia. Recent studies have explored the bidirectional relationship between disrupted sleep and these conditions. Sleep disorders can lead to chronic deprivation, which may exacerbate cognitive decline and mood disturbances. Investigating the precise mechanisms through which sleep disturbances influence depression and dementia may lead to strategies for improving sleep quality. [32]

Neuroimaging and Biomarker Discovery: Advances in neuroimaging techniques, functional magnetic resonance imaging and positron emission tomography have allowed researchers to examine structural and functional changes in the brain linked to depression and dementia. These techniques allow for the identification of potential biomarkers that could serve as early indicators or targets for intervention. Ongoing investigations aim to pinpoint specific neuroimaging biomarkers that may help with early diagnosis and monitoring of disease progression. [33]

Psychosocial Factors: Beyond biological and physiological aspects, the impact of psychosocial factors on the depression-dementia link is gaining attention. Recent research has been able to discover the role of social isolation, loneliness, and socioeconomic status in influencing depression and dementia. Having a deep understanding of how psychosocial determinants interact with biological factors could inform comprehensive interventions such as addressing medical aspects, social, and emotional well-being of individuals. [34]

Lifestyle Interventions such as diet, physical activity, and cognitive stimulation are emerging as potential strategies for mitigating the risk of depression and dementia. Research on the impact of dietary patterns such as the Mediterranean diet and regular physical exercise on cognitive function and mood is ongoing. Cognitive training programs such as mental stimulation activities are being investigated for their potential to preserve cognitive abilities and reduce the risk of dementia. [35]

A shift towards precise medicine is becoming evident such as tailoring interventions based on an individual's genetics, molecular, and clinical profile holds promise in optimizing treatment and prevention strategies. Precise medicine will allow healthcare providers to identify personalized risk factors and develop targeted interventions, which will improve outcomes for individuals at risk of depression and dementia. [36]

Part 4: Case Studies on Alzheimer's

Case Study 1

This case study highlights a 63-year-old woman white female who had completed 2 years of postsecondary education and worked as a manager in a telecommunications company until her retirement at age 48. The patient suffered from early-onset Alzheimer’s disease (EOAD). Early onset Alzheimer’s disease (EOAD) is a rare form of neurodegenerative dementia that affects less than 1% of all Alzheimer’s cases. The degeneration of brain tissue and cognitive decline classifies this condition. Genetic mutations in the presenilin 1 (PSEN1), presenilin 2 (PSEN2), and amyloid precursor protein (APP) genes have been associated with early-onset Alzheimer's disease. Mutations in PSEN1 and PSEN2 genes can lead to altered activity of the γ-secretase enzyme, which leads to increased production of the amyloidogenic Aβ42 protein, formation of amyloid plaques.

The patient displayed various symptoms which included memory loss, executive functions such as decision and problem solving, and difficulties with word finding and literacy skills.

In the beginning stage, her symptoms were characterized by repetitive memory loss, a loss of executive functions, deterioration in her ability to find words and comprehend written material, and slow execution of instruction. The daily life problems the patient was facing include difficulty turning off stoves, an inability to assist in boat docking, poor verbal fluency, poor visuospatial skills, word finding difficulties, and mid ideomotor apraxia. The patient's speech remained fluent without semantic deficits, and she didn't make phonemic or semantic paraphasic errors. She exhibited normal muscle tone and power, but faced mild memory loss. Hypercholesterolemia known as having high cholesterol levels and a vitamin D deficiency were noted. Her condition worsened, as time went on, leading to a decline in verbal fluency, visuospatial skills, and memory.

Understanding the patient's family history is important in assessing potential risk factors for Alzheimer's disease. Her mother, who had a history of long-term smoking and idiopathic pulmonary fibrosis known as scarring of lung tissue with no known cause, passed away at the age of 79 due to a hip fracture. Long-term smoking is considered a risk factor for Alzheimer's disease due to its contribution to oxidative stress and inflammation. There is no direct link that exists between idiopathic pulmonary fibrosis and Alzheimer's disease, chronic lung diseases have been associated with cognitive decline. The patient's father, who had a history of coronary artery disease, was alive and well during the patient's presentation. Coronary artery disease is known to be associated with an increased risk of Alzheimer's disease, due to the shared genetic and environmental factors of cardiovascular health.The patient's paternal grandfather died at a young age due to appendicitis. Her paternal grandmother experienced mild memory loss without known dementia until the age of 76.Mild memory loss could indicate early cognitive changes however it is not possible to establish a direct link to Alzheimer's disease. The patient's paternal uncle was diagnosed with Parkison's disease in his 40s and passed away at 58 years old. Parkison's disease is a neurological disorder and can coexist with Alzheimer's disease. Her maternal grandparents had multiple health problems such as myocardial infarction and congestive heart failure which are associated with the increase of Alzheimer's disease. The patient had three siblings ranging from 44 to 56 years old and none of whom reported any issues associated with cognitive problems at the time of presentation.

The patient received a diagnosis of early-onset Alzehimer's disease based on clinical evaluation. Other diagnoses considered were the posterior cortical variant of Alzheimer's disease, atypical presentation of frontotemporal dementias, and Lewy Body dementia. The patient was prescribed a cholinesterase inhibitor and memantine medications for Alzheimer's disease.

As the disease progressed, the patient experienced a decline in cognition and function, necessitating admission to a care facility and complete dependence on others for basic activities of daily living. Postmortem examination confirmed the presence of Alzheimer's disease pathology in the patient's brain, characterized by the presence of plaques and tangles, congophilic amyloid angiopathy, and Lewy Body pathology. The findings were consistent with the clinical diagnosis of early-onset Alzheimer's disease.

Her medical genetic testing revealed a novel PSEN2 mutation that had not been reported before. The mutation involved a change in the DNA sequence at codon 222, resulting in the substitution of adenosine for cytosine and lysine replacing threonine. The patient was found to be a heterozygote for the apoE4 allele. Genetic mutations in PSEN1 and PSEN2 genes have been linked to familial forms of Alzheimer's disease. The pathogenicity of the novel PSEN2 mutation identified in this patient (L221T) requires further investigation to determine its specific role in disease progression and manifestation. The apoE4 allele is a well-known genetic risk factor for late-onset Alzheimer's disease. Its association with early-onset cases is less clear but suggests a possible increased risk in this patient. [37]

Case Study 2

The individual under examination is a 63-year-old man, married and engaged in business, hailing from Shanghai. His symptoms encompass a range of distressing experiences, including short-term memory loss, difficulties in expressing himself, becoming socially withdrawn, undergoing shifts in his personality, struggling with clear speech, grappling with forgetfulness, encountering issues with language usage, constructing sentences with difficulty, finding it hard to recognize his own reflection, hiding personal possessions out of fear of theft, experiencing volatile moods, frequently breaking into tears, facing challenges in personal care, battling depression, encountering visual hallucinations, being irritable, and facing deficits in various cognitive abilities such as orientation, attention, calculation, recall, and visual tasks. His past medical history points to a recovery from hepatitis A over three decades ago and well-managed hypertension for thirty years.

The diagnostic procedures carried out unveil critical insights into his condition. Brain MRI revealed a progressive atrophy (shrinkage) of the bilateral frontal lobes and, subsequently, the entire cerebral cortex. FDG-PET scans (involves injecting a mildly radioactive substance similar to glucose into the body, and the scanner then detects areas with higher glucose metabolism, helping visualize brain activity and identifying changes in brain function) showed reduced glucose metabolism in the frontal and parietal lobes of his brain, indicative of reduced brain activity in those areas. His Mini-Mental State Examination (MMSE) scores, a common cognitive assessment tool, declined to as low as 3 out of 30. The MMSE is a test used by doctors to measure a person's cognitive (thinking and memory) abilities. A score of 30 is perfect, indicating normal cognitive function. In this case, the individual's MMSE score dropped dramatically to just 3 out of 30, suggesting severe cognitive impairment. Cerebrospinal fluid (CSF) analysis indicated decreased levels of a protein called amyloid β-protein (Aβ) 42, and elevated levels of other proteins called total tau and phosphorylated tau, which are associated with brain cell damage. A more advanced imaging technique known as 11C-Pittsburgh compound B positron emission tomography (11C-PIB PET) confirmed the presence of extensive deposits of amyloid β-protein in various regions of the brain, particularly in areas responsible for thinking and memory. Genetic testing did not find any mutations in genes that are typically linked to early-onset Alzheimer's disease (EOAD), but it did identify a specific genetic type called Apolipoprotein E (APOE), more specifically APOE ε3/ε3 type.

The patient's treatment journey involved medications like rivastigmine, quetiapine, donepezil, and memantine, aimed at managing memory, hallucinations, cognition, and mood. The patient was also provided with non-drug interventions and psychosocial support, which included therapies and support for both the patient and his caregiver. Ultimately, the diagnosis settled on a particular subtype of early-onset Alzheimer's disease known as frontal variant early-onset Alzheimer's disease (fvAD). This was backed by confirmatory evidence from the 11C-PIB PET scan and CSF biomarkers. Notably, the complexity of the case suggests a mix of neuropathologies, potentially involving both frontotemporal lobar degeneration and Alzheimer's disease. In essence, this comprehensive examination demonstrates the intricate interplay between cognitive symptoms, neurological changes, and genetic factors in Alzheimer's spectrum disorders, highlighting the multifaceted nature of the disease.

The patient's treatment journey involved medications such as rivastigmine, quetiapine, donepezil, and memantine, targeting memory, hallucinations, cognition, and mood. Non-drug interventions and psychosocial support were also provided for both the patient and his caregiver. Ultimately, the diagnosis settled on frontal variant early-onset Alzheimer's disease (fvAD), supported by evidence from the 11C-PIB PET scan and CSF biomarkers. Notably, the complexity of the case suggests a potential mix of neuropathologies, possibly involving both frontotemporal lobar degeneration and Alzheimer's disease. This comprehensive examination underscores the intricate interplay between cognitive symptoms, neurological changes, and genetic factors in Alzheimer's spectrum disorders, highlighting the multifaceted nature of the disease. [38]

Case Study 3

This case study is about a 37-year-old guy who had early-onset dementia. Numerous symptoms were present in the patient, including math errors, frequent amnesia, apraxia, apathy, confusion to time and person, and night chatting. The symptoms began three years ago. The patient worked as an industrial researcher before retiring owing to memory loss.

Prior to attending the specific clinic, he went to two hospitals for diagnosis and therapy of his symptoms since memory difficulties prohibited him from performing his job obligations. However, the results were normal. After a comprehensive assessment of his symptoms revealed no anomalies, the patient was admitted to a psychiatric institution. His tests for the human immunodeficiency virus came back negative. To assess his cognitive status, a battery of neuropsychological tests was administered. He received a 22 on the Mini-mental Status Exam. He received a score of 1 on the Clinical Dementia Rating Scale (CDR). He received a 4.5 on cognitive testing and had significant difficulties on free calls. He also exhibited a delayed recall and recognition of twenty minutes. He had a cerebral magnetic resonance imaging scan.

He was given 5 mg of donepezil and was discharged on the 10th day of his stay. Donepezil is a cholinesterase inhibitor. It increases mental skills such as memory, attention, and the capacity to interact with others, talk, and conduct daily activities. They prevent acetylcholine from being broken down, which alleviates dementia symptoms. Donepezil was raised to 23 mg with 20 mg memantine to control his cognitive impairment after discharge. Memantine works by blocking the effects of glutamate. Glutamate communicates between nerve cells in the brain. However, Alzheimer's disease produces an excessive amount of glutamate, which damages nerve cells. Finally, memantine inhibits the consequences of excess glutamate. The patient must now return to the clinic on a regular basis. Early onset dementia, criterion 16 for probable Alzheimer's disease, and early onset of cognitive impairment with fast deterioration are all officially identified in the patient. [39]

Case Study 4

This case study highlights Mr.A, a 68-year-old male engineer who experienced a variety of symptoms over a decade. Mr.A was found to be suffering from Alzheimer's disease after extensive diagnostic tests revealed that the illness ran in his family. This case study considers his medical background, family history, diagnostic evaluations, and the neuropathological signs that led to 898926- the final diagnosis. The results highlight the significant role that familial predisposition plays in the development of Alzheimer's disease and highlight its devastating effects.

Mr.A, a 68-year-old man marked by cognitive decline, atypical language use, and severe memory loss. His story illuminates the clinical signs and familial implications of Alzheimer's Disease, offering no insight into the complex network of the disease. Mr.A is a married engineer with 16 years of education, displaying verbal communication issues, finding it difficult to form his sentences when speaking. His verbal speech was characterized by frequent word changes and inability to remember names. He also battled dysgraphia, which caused a decrease in his writing skills and showed cognitive deterioration, getting angrier and losing interest in new activities. Over the decade, these symptoms worsened.

According to Mr.A's medical history, he has a past of having high blood pressure that has required medication. He had unrelated stomach and foot surgeries a decade ago. These surgical procedures were deemed unrelated to his present state of cognition.

Moving on to his family history, noted that Mr.A's mother passed away from dementia, Alzheimer's disease was an important aspect of his family history. Important contextual information included his father's heart attack death at age 79 and the absence of mental disorders in his two older siblings.

Diagnostic evaluations revealed the severity of Mr.A's condition. Neurological tests revealed modest attentional and focus issues as well as the inability to correctly spell a five letter word backwards. Paraphrasing and anomia mistakes were among the more noticeable language problems. Results from physical exams, blood tests, a cerebrospinal fluid analysis, and chest X-rays were all normal. A brain scan revealed slow brain active, the results of neuropsychological testing revealed serious deficiencies in arithmetic skills as well as memory and language functions. Mr.A's brain showed signs of shrinkage and enlargement of fluid-filled cavities after his postmortem examination. amyloid plaques and neurofibrillary tangle, the hallmarks of Alzheimer's disease were present in the screening.

Mr.A's case resulted in a diagnosis of Alzheimer's disease. He displayed Parkinson's-like symptoms initially, and further examination indicated that his ailment was postsynaptic and caused by injury to the cerebral cortex and striatum.

Mr.A is a good example of unrelenting advancement of Alzheimer's disease. The fact that this crippling ailment runs in families underscores how important genetic predisposition was in its genesis. This case study helps us comprehend Alzheimer's disease by its nuanced symptoms and the significant toll it takes on sufferers and their families. [40]

Case Study 5

This case study is about a 19-year old man patient who was described with high chances of Alzheimer's disease. The patient had no known genetic markers for Alzheimer's; however, the patient displayed clinical symptoms and physiological signs. Hence, this example sheds light on the rising occurrence of early-onset Alzheimer's in young adults.

People under the age of 65 who have Alzheimer's disease is a very uncommon illness that only accounts for 5% to 10% of all Alzheimer's cases. With the recent case study presenting a 19-year old patient's diagnosis of high chances of Alzheimer's disease, it has brought more attention on preventing cognitive decline.

There were no valid genetic abnormalities in the 19 year-old patient, which calls into question the understanding of the disease, since early-onset Alzheimer's disease for those under 30 is connected to certain gene alterations. This case study highlights the variety of symptoms that can lead to dementia and demonstrates a need for a reevaluation of Alzheimer's disease for the different age groups.

This specific case of early-onset Alzheimer's disease demonstrates the disorder occurring in younger population. This case study shows that early-onset Alzhimer's is complex, its not just about genetics, lifestyle, and environmental variables. The need for more Alzhimer's disease research and development of treatment options for individuals that are experiencing early-onset Alzheimer's at a younger age should be more looked into. [41]

Case Study 6

A 17-year old male patient from Beijing, China was experiencing memory loss and lack of focus. These symptoms over time can cause issues with memory recall and cognitive decline. After a neurocognitive evaluation, including the WHO-UCLA AVLT, revealed substantial memory impairment. The following results were discovered after medical investigations:

There was bilateral hippocampal shrinkage and hypometabolism in the contralateral temporal lobe after a neuroimaging with 18F fluorodeoxyglucose. These signs are recognized as early signs of Alzheimer's disease. With the addition of a cerebrospinal fluid examination, it indicated an enhanced p-tau181 concentration and a reduced amyloid-42/40 ratio which confirmed the diagnosis. However, the whole-genome sequencing revealed no valid gene alterations linked to Alzheimer's disease.

SUMMARY

NMN and other Anti-Aging Supplements:

Anti-aging supplements, a novel field which have been developed in recent years, decrease the harmful effects of free radicals, molecules which can cause disease. Vitamin C, which promotes collagen synthesis to decrease wrinkles, and Vitamin E, which protects skin against UV radiation, are natural anti-aging supplements that are both found in plants. Research in the field of anti-aging is supported by research on induced pluripotent stem cells and tissue repair for muscle dystrophy.

One prominent anti-aging supplement is resveratrol, a natural substance found in the root of the white hellebore plant in 1939. Resveratrol is able to inhibit cancer through its many properties, such as reaching different molecular targets without toxicity. Another property includes regulating genes that affect cancer cell proliferation. Furthermore, resveratrol induces apoptosis through certain pathways such as the PI3K/Akt/mTOR pathway. Also, resveratrol can inhibit oxidative stress through targeting free radicals created by cells. Lastly, resveratrol uses its anti-inflammatory effects to affect diseases such as Crohn's Disease, an inflammatory disease in the bowel area.

However, resveratrol has a few drawbacks such as decreased bioavailability and solubility, causing the drug to be less effective. Research has been done to increase resveratrol bioavailability through taking resveratrol with flavonoids and cyclodextrins, two other supplements. In addition, resveratrol bioavailability can be increased through nanoization, which increases surface area for absorption.

In multiple clinical trials, resveratrol has been researched to have anti-aging effects, including improving memory, protecting the cardiovascular system, and being a deterrent to age-induced diabetes. More research needs to be done on resveratrol to determine its complete effects on anti-aging.

Besides resveratrol, Nicotinamide Mononucleotide (NMN) is another prominent anti-aging supplement. The authors of this chapter conducted a human clinical trial to measure the effects of NMN on feelings of youthfulness. A 10-question questionnaire was created based on parameters of youthfulness, such as energy level and happiness. 16 human subjects were recruited, with 12 subjects being experimental and 4 being control. The experimental human subjects each took one NMN capsule every day for 3 months, while the control human subjects took placebo pills filled with protein powder. In the data analyzed, it reveals that there is a greater average increase in feelings of youthfulness for the experimental subjects compared to the control subjects.

Furthermore, NMN has been tested in multiple clinical trials and shown positive effects in reducing obesity in older adults, a higher blood NAD level, and showed greater oxygen intake. In the future, NMN could be used to improve insulin sensitivity and benefit Alzheimer's Disease patients.

Veganism and Cardiovascular Disease:

It's crucial to remember that every person reacts differently to a vegan diet and the diet's overall quality counts. Cardiovascular health can benefit from a well-planned, well-balanced vegan diet that satisfies nutritional demands; nevertheless, to guarantee adequate nutrition and address specific health issues, it is best to speak with a healthcare provider or certified dietitian.

Diets mainly based on plants often include less saturated fat, primarily found in animal products. Reducing saturated fat consumption can help decrease LDL ("bad") cholesterol, lowering the risk of cardiovascular disease and atherosclerosis. Blood pressure is frequently found to be reduced by vegan diets. Consuming a lot of fruits, vegetables, healthy grains, legumes, and less salt may help regulate blood pressure.

Vegan diets may help with insulin sensitivity and diabetes management. Reducing the risk of diabetes-related cardiovascular problems requires improved blood sugar regulation.

Plant fiber is often abundant in plant-based diets, which can help maintain a healthy weight, control blood sugar, and decrease cholesterol. A lower risk of heart disease is linked to adequate fiber consumption.

Antioxidants are abundant in fruits, vegetables, nuts, and seeds—essential foods for a vegan diet. Through their ability to neutralize free radicals, reduce inflammation, and stop oxidative stress, antioxidants assist in protecting the cardiovascular system.

Changing to a vegan diet might help you lose or keep weight. Since obesity raises the risk of heart disease, maintaining a healthy weight is crucial for cardiovascular health.

Diets based chiefly on plants have been linked to decreased inflammatory levels in the body. All the necessary nutrients may be obtained with a well-planned vegan diet. However, particular nutrients, such as vitamin B12, iron, calcium, omega-3 fatty acids, and vitamin D, need to be considered explicitly by individuals. Inadequate intake of certain nutrients may affect cardiovascular health.

While there are many types of diets, a vegan diet has shown to have an abundance of health benefits greater in variety than others. Some benefits of adopting a vegetarian diet include reduced risk of heart disease, cancer, and joint pain caused by arthritis. While many are concerned about the lack of protein that may result of adopting a vegan diet, there are many options that can be selected such as lentils, beans, and peas. LDL cholesterol among other risk factors for cardiovascular disease can be limited with the help of a vegan diet.

Veganism and Diabetes:

The global rise in Type 2 Diabetes means that a thorough understanding of risk factors, complications, and management strategies is necessary. This chapter explores the multifaceted dangers of diabetes and demonstrates its impact on many aspects of human health. A proposed solution to diabetes, which is a plant-based diet, is supported by a meta-analysis of three research studies, as it shows a great reduction in the risk of Type 2 Diabetes among research subjects adhering to a plant-based diet.

The research of this chapter analyzes participants from Taiwan and North America, carefully categorizing them into plant-based and non-plant-based groups. Using chi-squared analysis, the data shows a statistically significant difference in the rates of Type 2 Diabetes in plant-based and non-plant-based groups, demonstrating the potential impact of dietary choices on preventing diabetes.

The research of this chapter explores the broader advantages of a plant-based diet beyond just the impact on diabetes, including cardiovascular well-being, increased energy, improved digestive health, weight management, and positive environmental impact. This chapter highlights the lifestyle impacts of adopting a plant-based approach, not just as a dietary choice but as a way to enhance a person's well-being.

Besides pointing out limitations in the research, such as the need for more diverse data across the world, the research spotlights the importance of controlling for multiple factors in the selected studies. In addition, it advocates for current research on multiple facets of the connection between veganism and diabetes management, including the factors of glycemic control, weight management, cardiovascular health, gut microbiota, long-term outcomes, behavioral aspects, and comparative effectiveness.

To conclude, the comprehensive analysis presented in this exploration demonstrates a convincing connection between a plant-based diet and a reduced risk of Type 2 Diabetes. As scientists continue to discover the nuances of this relationship, patients are encouraged to consider the potential benefits of embracing a plant-based diet for a healthier lifestyle.

Veganism and Cancer:

Cancer in relation to diet is mainly characterized by the following internal organs : the stomach, esophagus, pancreas, kidney, gallbladder, and rectum. Despite being very hard to monitor and consequently, diagnose any sort of diseases, there are many steps to take that can lessen the risk factors involved.

In America, nearly 42% of adults have obesity, all of whom are therefore placed in a higher risk category for the cancers listed above. The so called "obesity epidemic" is caused by unhealthy diets, consisting of fatty and sugary foods, combined with a lack of exercise. In fact, around half of Americans do not have a "healthy" (consisting of foods such as many vegetables, fruits, whole grains, etc.) diet, and around 40% do not exercise in a moderate way. All of these factors contribute to the perpetration of various cancers.

The chapter will discuss findings in studies relating to various types of diets : mediterranean, vegetarian, vegan. Although one may be quick to associate certain foods with being healthy and maybe even cancer fighting, they may not know the actual causes. To provide some reasoning, which only scrapes the surface of what is discussed in this chapter, we can take a look at "antioxidants". Many people have heard this word and may associate it with foods like grapes or blueberries, and even types of fish and carrots. The basis for these foods being "cancer-fighting" lies in their chemical properties. These foods harbor a variety of compounds that are classified as antioxidants, and these compounds are observed to "attack" free radicals, neutralizing their charge. This is important as free-radicals are known to be highly concentrated in tumors. Without delving too much into the topic, since these free radicals are being neutralized, the anti-oxidants are almost attacking the cancer, fighting it off.

Although there is still much to be investigated about the correspondence between food and cancer, the studies that are discussed late in the chapter will provide the logic and reasoning behind a diet's effect on cancer.

Alzheimer's Disease:

Alzheimer's disease, which is an irreversible neurodegenerative disorder with characteristics of lessened memory, impacts millions globally, with its main cause in genetics, lifestyle factors, and the environment. The pathology of Alzheimer's Disease includes the increase of abnormal protein in the brain, inducing inflammation and cellular damage, and affects women more than men. Possible mitigations of the disease include changing to healthier diets like the Mediterranean diet. Current studies are exploring the impact of gut health, sleep patterns, and stress on how the brain functions. This is done to pinpoint dietary and lifestyle interventions for improving memory, seeking to make treatments better and alleviate the societal damage of Alzheimer's, as well as depressive disorders.

The five case studies of this chapter contain valuable insights into how Alzheimer's Disease can affect people and how the disease is diagnosed. In Case Study 1, a 63-year-old woman has a rare form of Alzheimer's, which emphasizes the role of genetics in development of Alzheimer's. In Case Study 2, a 63-year-old man has multiple Alzheimer's symptoms, which demonstrates the complications of the disease. In Case Study 3, a 37-year-old man had trouble with his memory, highlighting the challenges of an Alzheimer's diagnosis. In Case Study 4, a 68-year-old engineer has Alzheimer's and has relatives with Alzheimer's, showing how family history can play a role in Alzheimer's. Finally, in Case Study 5, a 19-year-old has Alzheimer's, which is surprising for such a young patient. These stories help us understand that Alzheimer's disease can affect individuals in multiple ways, and how it is much more than just a disease of old age. Family history, genetics, and other factors all play a role in how Alzheimer's Disease develops.

SOURCES

Chapter 1: NMN and Other Supplements

[1] Arora, Balvant P. “Anti-Aging Medicine.” *Indian Journal of Plastic Surgery : Official Publication of the Association of Plastic Surgeons of India*, vol. 41, no. Suppl, 1 Oct. 2008, pp. S130–S133, www.ncbi.nlm.nih.gov/pmc/articles/PMC2825135/.

[2] Better Health Channel. “Antioxidants.” *Vic.gov.au*, 30 June 2022, www.betterhealth.vic.gov.au/health/healthyliving/antioxidants.

[3] Harvard School of Public Health. “Vitamin C.” *The Nutrition Source*, Harvard, Mar. 2023, www.hsph.harvard.edu/nutritionsource/vitamin-c/.

[4] Barney, Alicia. “The Benefits of Vitamin C for Your Skin.” *WebMD*, www.webmd.com/beauty/ss/slideshow-benefits-of-vitamin-c-for-skin#:~:text=Vitamin%20C%20is%20a%20powerful. Accessed 1 July 2024.

[5] “Vitamin E and Skin Health.” *Linus Pauling Institute*, 7 Nov. 2016, lpi.oregonstate.edu/mic/health-disease/skin-health/vitamin-E#:~:text=Vitamin%20E%20has%20been%20considered.

[6] “3 Benefits of Vitamin E for Skin and How to Use It.” *Ro*, ro.co/health-guide/vitamin-e-for-skin/#:~:text=Vitamin%20E%20can%20help%20keep. Accessed 1 July 2024.

[7] “The Truth behind Anti-Aging Supplements, Vitamins, and Pills.” *Blog.insidetracker.com*, blog.insidetracker.com/anti-aging-supplements-pills.

[8] “The Nobel Prize in Physiology or Medicine 2012.” *NobelPrize.org*, 2012, www.nobelprize.org/prizes/medicine/2012/yamanaka/facts/.

[9] *Gladstone.org*, gladstone.org/people/shinya-yamanaka.

[10] “Shinya Yamanaka | Japanese Scientist.” *Encyclopedia Britannica*, www.britannica.com/biography/Shinya-Yamanaka.

[11] “Thomas A. Rando, M.D., Ph.D. | UCLA BSCRC.” *Stemcell.ucla.edu*, stemcell.ucla.edu/member/rando. Accessed 1 July 2024.

[12] “Dr. Thomas Rando Receives Prestigious NOMIS Foundation Award to Study Stem Cell Quiescence.” *Www.uclahealth.org*, www.uclahealth.org/news/RandoNOMIS2022. Accessed 1 July 2024.

[13] “Donn Young, PhD | OSUCCC – James Patient.” *The James - OSUCCC*, cancer.osu.edu/for-patients-and-caregivers/learn-about-cancers-and-treatments/innovation-at-the-james/clinical-trials/clinical-trial-patient-stories/donn-young. Accessed 9 Dec. 2023.

[14] “Prostate Cancer - Statistics.” *Cancer.net*, 25 June 2012, www.cancer.net/cancer-types/prostate-cancer/statistics#:~:text=The%205%2Dyear%20relative%20survival%20rate%20for%20prostate%20cancer%20in.

[15] Bulterijs, Sven, et al. “It Is Time to Classify Biological Aging as a Disease.” *Frontiers in Genetics*, vol. 6, 18 June 2015, www.ncbi.nlm.nih.gov/pmc/articles/PMC4471741/, https://doi.org/10.3389/fgene.2015.00205.

[16] Pezzuto, John M. “Resveratrol: Twenty Years of Growth, Development and Controversy.” *Biomolecules & Therapeutics*, vol. 27, no. 1, 1 Jan. 2019, pp. 1–14, www.ncbi.nlm.nih.gov/pmc/articles/PMC6319551/, https://doi.org/10.4062/biomolther.2018.176.

[17] Ko, Jeong-Hyeon, et al. “The Role of Resveratrol in Cancer Therapy.” *International Journal of Molecular Sciences*, vol. 18, no. 12, 1 Dec. 2017, www.ncbi.nlm.nih.gov/pubmed/29194365, https://doi.org/10.3390/ijms18122589. Accessed 8 December 2023.

[18]: Salehi, Bahare, et al. “Resveratrol: A Double-Edged Sword in Health Benefits.” *Biomedicines*, vol. 6, no. 3, 9 Sept. 2018, p. 91, www.ncbi.nlm.nih.gov/pmc/articles/PMC6164842/#B46-biomedicines-06-00091, https://doi.org/10.3390/biomedicines6030091.

[19] García, Patricia, et al. "Resveratrol Causes Cell Cycle Arrest, Decreased Collagen Synthesis, and Apoptosis in Rat Intestinal Smooth Muscle Cells." *American Journal of Physiology-Gastrointestinal and Liver Physiology*, vol. 302, no. 3, 1 Feb. 2012, pp. G326–G335, https://doi.org/10.1152/ajpgi.00083.2011. Accessed 19 Aug. 2023.

[20] Zeng, Guofang et al. "Resveratrol-mediated reduction of collagen by inhibiting proliferation and producing apoptosis in human hypertrophic scar fibroblasts." *Bioscience, biotechnology, and biochemistry* vol. 77,12 (2013): 2389-96. doi:10.1271/bbb.130502

[21] Almatroodi, Saleh A et al. "Potential Therapeutic Targets of Resveratrol, a Plant Polyphenol, and Its Role in the Therapy of Various Types of Cancer." *Molecules (Basel, Switzerland)* vol. 27,9 2665. 21 Apr. 2022, doi:10.3390/molecules27092665

[22] Singh, Santosh Kumar et al. "Resveratrol induces cell cycle arrest and apoptosis with docetaxel in prostate cancer cells via a p53/ p21WAF1/CIP1 and p27KIP1 pathway." *Oncotarget* vol. 8,10 (2017): 17216-17228. doi:10.18632/oncotarget.15303

[23] Pannu, Naveet, and Archana Bhatnagar. "Resveratrol: From Enhanced Biosynthesis and Bioavailability to Multitargeting Chronic Diseases." *Biomedicine & Pharmacotherapy*, vol. 109, Jan. 2019, pp. 2237–2251, https://doi.org/10.1016/j.biopha.2018.11.075. Accessed 8 December 2023.

[24] Branton, Edgar. "Resveratrol Delivery." *News-Medical.net*, 4 Mar. 2010, www.news-medical.net/health/Resveratrol-Delivery.aspx#:~:text=Resveratrol%20administered%20in%20combination%20with. Accessed 9 Dec. 2023.

[25] Witte, A. V., et al. "Effects of Resveratrol on Memory Performance, Hippocampal Functional Connectivity, and Glucose Metabolism in Healthy Older Adults." *Journal of Neuroscience*, vol. 34, no. 23, 4 June 2014, pp. 7862–7870, https://doi.org/10.1523/jneurosci.0385-14.2014. Accessed 28 Feb. 2019.

[26] Ghinda, Diana C., et al. "How Much Is Enough—Can Resting State FMRI Provide a Demarcation for Neurosurgical Resection in Glioma?" *Neuroscience & Biobehavioral Reviews*, vol. 84, Jan. 2018, pp. 245–

261, https://doi.org/10.1016/j.neubiorev.2017.11.019. Accessed 19 Feb. 2022.

[27] Berman, Adi Y., et al. “The Therapeutic Potential of Resveratrol: A Review of Clinical Trials.” *Npj Precision Oncology*, vol. 1, no. 1, 25 Sept. 2017, www.nature.com/articles/s41698-017-0038-6, https://doi.org/10.1038/s41698-017-0038-6. Accessed 30 Dec. 2019.

[28] Mordarska, Katarzyna, and Małgorzata Godziejewska-Zawada. “Diabetes in the Elderly.” *Menopausal Review*, vol. 16, no. 2, 2017, pp. 38–43, www.ncbi.nlm.nih.gov/pmc/articles/PMC5509969/, https://doi.org/10.5114/pm.2017.68589.

[29] Crandall, Jill P., et al. “Pilot Study of Resveratrol in Older Adults with Impaired Glucose Tolerance.” *The Journals of Gerontology: Series A*, vol. 67, no. 12, 4 Jan. 2012, pp. 1307–1312, https://doi.org/10.1093/gerona/glr235. Accessed 8 Dec. 2023.

[30] Centers for Disease Control and Prevention. “FastStats - Older Persons Health.” *Center for Disease Control and Prevention*, 2019, www.cdc.gov/nchs/fastats/older-american-health.htm.

[31] Sinclair, D.A., and LaPlante, M.D. *Lifespan : Why We Age--and Why We Don’t Have To*. New York, Atria Books, 2019.

[32] Shade, Christopher. “The Science behind NMN-A Stable, Reliable NAD+Activator and Anti-Aging Molecule.” *Integrative Medicine*, vol. 19, no. 1, 1 Feb. 2020, pp. 12–14.

[33] Yoshino, Mihoko, et al. “Nicotinamide Mononucleotide Increases Muscle Insulin Sensitivity in Prediabetic Women.” *Science*, vol. 372, no. 6547, 22 Apr. 2021, science.sciencemag.org/content/early/2021/04/21/science.abe9985, https://doi.org/10.1126/science.abe9985.

[34] Nadeeshani, Harshani, et al. “Nicotinamide Mononucleotide (NMN) as an Anti-Aging Health Product – Promises and Safety Concerns.” *Journal of Advanced Research*, vol. 37, no. 37, 1 Mar. 2022, pp. 267–278, www.sciencedirect.com/science/article/pii/S2090123221001491#:~:text =Among%20various%20anti-

aging%20health%20products%2C%20nicotinamide%20mononucleotide%20%28NMN%29, https://doi.org/10.1016/j.jare.2021.08.003.

[35] Long, Aaron N., et al. "Effect of Nicotinamide Mononucleotide on Brain Mitochondrial Respiratory Deficits in an Alzheimer's Disease-Relevant Murine Model." *BMC Neurology*, vol. 15, 1 Mar. 2015, p. 19, pubmed.ncbi.nlm.nih.gov/25884176/, https://doi.org/10.1186/s12883-015-0272-x. Accessed 14 Nov. 2022.

[36]Yi, Lin et al. "The efficacy and safety of β-nicotinamide mononucleotide (NMN) supplementation in healthy middle-aged adults: a randomized, multicenter, double-blind, placebo-controlled, parallel-group, dose-dependent clinical trial." *GeroScience* vol. 45,1 (2023): 29-43. doi:10.1007/s11357-022-00705-1

[37] Pencina, Karol M et al. "MIB-626, an Oral Formulation of a Microcrystalline Unique Polymorph of β-Nicotinamide Mononucleotide, Increases Circulating Nicotinamide Adenine Dinucleotide and its Metabolome in Middle-Aged and Older Adults." *The journals of gerontology. Series A, Biological sciences and medical sciences* vol. 78,1 (2023): 90-96. doi:10.1093/gerona/glac049

[38] Katayoshi, Takeshi et al. "Nicotinamide adenine dinucleotide metabolism and arterial stiffness after long-term nicotinamide mononucleotide supplementation: a randomized, double-blind, placebo-controlled trial." *Scientific reports* vol. 13,1 2786. 16 Feb. 2023, doi:10.1038/s41598-023-29787-3

[39] Xiao, S., Chan, P., Wang, T. *et al.* A 36-week multicenter, randomized, double-blind, placebo-controlled, parallel-group, phase 3 clinical trial of sodium oligomannate for mild-to-moderate Alzheimer's dementia. *Alz Res Therapy* 13, 62 (2021). https://doi.org/10.1186/s13195-021-00795-7

[40] Harshani Nadeeshani, et al. "Nicotinamide Mononucleotide (NMN) as an Anti-Aging Health Product – Promises and Safety Concerns." *Journal of Advanced Research*, Elsevier, 11 Aug. 2021, www.sciencedirect.com/science/article/pii/S2090123221001491.

[41] Hong, Weiqi et al. "Nicotinamide Mononucleotide: A Promising Molecule for Therapy of Diverse Diseases by Targeting NAD+

Metabolism." *Frontiers in cell and developmental biology* vol. 8 246. 28 Apr. 2020, doi:10.3389/fcell.2020.00246

[42] Mihoko Yoshino *et al.*,Nicotinamide mononucleotide increases muscle insulin sensitivity in prediabetic women.Science 372,1224-1229(2021).DOI:10.1126/science.abe9985

Chapter 2: Veganism and Cardiovascular Disease

[1] Roman Pawlak PhD, et al. "Is Vitamin B12 Deficiency a Risk Factor for Cardiovascular Disease in Vegetarians?" *American Journal of Preventive Medicine*, Elsevier, 18 May 2015, www.sciencedirect.com/science/article/abs/pii/S0749379715000732.

[2] *Plant-based Diets Are Associated with a Lower Risk of Incident ...*, www.ahajournals.org/doi/10.1161/JAHA.119.012865. Accessed 20 Sept. 2023.

[3] "Can a Plant-Based Diet 'reverse' Heart Disease?" *BHF*, British Heart Foundation, www.bhf.org.uk/informationsupport/heart-matters-magazine/nutrition/ask-the-expert/plant-based-diets. Accessed 20 Sept. 2023.

[4] Fanny Petermann-Rocha a b, et al. "Do All Vegetarians Have a Lower Cardiovascular Risk? A Prospective Study." *Clinical Nutrition*, Churchill Livingstone, 12 Jan. 2023, www.sciencedirect.com/science/article/pii/S0261561423000122.

[5] *Academic.Oup.Com*, academic.oup.com/eurjpc/article/29/14/1895/6612717. Accessed 20 Sept. 2023.

[6] *Dietary Supplementation with Lauric Acid Improves Aerobic Endurance In ...*, jn.nutrition.org/article/S0022-3166(23)72600-X/fulltext. Accessed 20 Sept. 2023.

[7]"Global Epidemiology of Obesity, Vegetarian Dietary Patterns, and Noncommunicable Disease in Asian Indians." *The American Journal of Clinical Nutrition*, Elsevier, 21 May 2014, www.sciencedirect.com/science/article/pii/S0002916523048827?via%3Dihub.

[8] *Vegetarian Diet and Incidence of Total, Ischemic, and ... - Neurology*, n.neurology.org/content/94/11/e1112. Accessed 9 Dec. 2023.

[9] Tong, Tammy Y N, et al. "Risks of Ischaemic Heart Disease and Stroke in Meat Eaters, Fish Eaters, and Vegetarians over 18 Years of Follow-up: Results from the Prospective Epic-Oxford Study." *The BMJ*, British Medical Journal Publishing Group, 4 Sept. 2019, www.bmj.com/content/366/bmj.l4897.

[10] Choi, Evan Y, et al. "A Plant-Based Diet and Heart Failure: Case Report and Literature Review." *Journal of Geriatric Cardiology : JGC*, U.S. National Library of Medicine, May 2017, www.ncbi.nlm.nih.gov/pmc/articles/PMC5466944/.

[11] Allen, Kathleen E., et al. "Heart Failure and a Plant-Based Diet. A Case-Report and Literature Review." *Frontiers*, Frontiers, 20 May 2019, www.frontiersin.org/articles/10.3389/fnut.2019.00082/full.

[12] Daniel, Diane. "After Surviving a Heart Attack at 35, She 'Felt like a Ticking Time Bomb.'" Www.Heart.Org, American Heart Association News, 2 June 2023, www.heart.org/en/news/2023/06/02/after-surviving-a-heart-attack-at-35-she-felt-like-a-ticking-time-bomb

[13] Bridgeman, Mary Barna, and Rupal Patel Mansukhani. "Case Studies: Cardiovascular Health." Pharmacy Times, MJH Life Sciences, 10 Jan. 2019, www.pharmacytimes.com/view/cardiovascular-health-.

[14] Mayo Clinic Staff. "How to Make the Dash Diet Work For You." Mayo Clinic, Mayo Foundation for Medical Education and Research, 25 May 2023, www.mayoclinic.org/healthy-lifestyle/nutrition-and-healthy-eating/in-depth/dash-diet/art-20048456.

[15] Zeratsky, Katherine. "Butter vs. Margarine: Which Is Better for My Heart?" Mayo Clinic, Mayo Foundation for Medical Education and Research, 1 Dec. 2022, www.mayoclinic.org/healthy-lifestyle/nutrition-and-healthy-eating/expert-answers/butter-vs-margarine/faq-20058152#:~:text=Margarine%20is%20a%20blend%20of%20oils%20that%20are%20mostly%20unsaturated%20fat.

[16] Editors of WebMD. "Plant Sterols: Overview, Uses, Side Effects, Precautions, Interactions, Dosing and Reviews." WebMD, WebMD, www.webmd.com/vitamins/ai/ingredientmono-1537/plant-sterols. Accessed 13 Dec. 2023.

[17] Jiang Z, Elkind M, Lucky M, et alStroke in a young man: a late complication of radiation therapyBMJ Case Reports CP 2019;12:e228029.

[18] Andersen MØ, Diederichsen SZ, Svendsen JH, et alContinuous cardiac monitoring in a patient with terminal pulmonary hypertension and eventual bilateral lung transplantationBMJ Case Reports CP 2022;15:e248304.

[19] Nunes RAB, Ramirez HSM, Aiello VD. Case 6 - Woman with Ischemic Heart Disease Admitted due to Chest Pain and Shock. Arq Bras Cardiol. 2018 Dec;111(6):860-863. doi: 10.5935/abc.20180231. PMID: 30517382; PMCID: PMC6263468.

[20] Shiba M, Kitano D, Kunimoto S, et alIdentification of left ventricular chamber-like aneurysm related to cardiac sarcoidosisCase Reports 2018;2018:bcr-2017-223910.

[21] Editors of The American Heart Association "The American Heart Association Diet and Lifestyle Recommendations." Www.Heart.Org, American Heart Association , 10 May 2023, www.heart.org/en/healthy-living/healthy-eating/eat-smart/nutrition-basics/aha-diet-and-lifestyle-recommendations.

Chapter 3: Veganism and Diabetes

[1] "Type 2 Diabetes." *Centers for Disease Control and Prevention*, Centers for Disease Control and Prevention, 30 Dec. 2022, https://www.cdc.gov/diabetes/basics/type2.html#:~:text=More%20than%2037%20million%20Americans,them%20have%20type%202%20diabetes.

[2] "Type 2 Diabetes." *Mayo Clinic*, Mayo Foundation for Medical Education and Research, 19 Nov. 2022, https://www.mayoclinic.org/diseases-conditions/type-2-diabetes/symptoms-causes/syc-20351193

[3] Katherine D. McManus, MS. "What Is a Plant-Based Diet and Why

Should You Try It?" *Harvard Health*, 16 Nov. 2021, https://www.health.harvard.edu/blog/what-is-a-plant-based-diet-and-why-should-you-try-it-2018092614760.
[4] MD Anderson Cancer Center, and Heather Alexander. "5 Benefits of a Plant-Based Diet." *MD Anderson Cancer Center*, MD Anderson Cancer Center, 4 Nov. 2020, www.mdanderson.org/publications/focused-on-health/5-benefits-of-a-plant-based-diet.h20-1592991.html#:~:text=It%20means%20your%20meals%20are,protein%20like%20tofu%20or%20beans.
[5] Tonstad, S, et al. "Vegetarian Diets and Incidence of Diabetes in the Adventist Health Study-2." *Nutrition, Metabolism, and Cardiovascular Diseases : NMCD*, U.S. National Library of Medicine, Apr. 2013, https://www.ncbi.nlm.nih.gov/pmc/articles/PMC3638849/.
[6] McMacken, Michelle, and Sapana Shah. "A Plant-Based Diet for the Prevention and Treatment of Type 2 Diabetes." *Journal of Geriatric Cardiology : JGC*, U.S. National Library of Medicine, May 2017, https://www.ncbi.nlm.nih.gov/pmc/articles/PMC5466941/.

[7] McMacken, Michelle, and Sapana Shah. "A Plant-Based Diet for the Prevention and Treatment of Type 2 Diabetes." *Journal of Geriatric Cardiology : JGC*, U.S. National Library of Medicine, May 2017, https://www.ncbi.nlm.nih.gov/pmc/articles/PMC5466941/#:~:text=Diets%20based%20in%20whole%20and,shown%20to%20promote%20insulin%20sensitivity.
[8] Dietitian, Plant Based. "Everything You Need to Know about a Vegan Diet in Less than 500 Words - Plant-Based Dietitian®." *Plant*, 13 Jan. 2023, https://plantbaseddietitian.com/vegan-diet-500-words/.
[9] Chiu, Tina H. T., et al. "Vegetarian Diet, Change in Dietary Patterns, and Diabetes Risk: A Prospective Study." *Nature News*, Nature Publishing Group, 9 Mar. 2018, https://www.nature.com/articles/s41387-018-0022-4.

[10] Dietitian, Plant Based. "Everything You Need to Know about a Vegan Diet in Less than 500 Words - Plant-Based Dietitian®." *Plant*, 13 Jan. 2023, https://plantbaseddietitian.com/vegan-diet-500-words/.

[11] "Fiber: The CARB That Helps You Manage Diabetes." *Centers for Disease Control and Prevention*, Centers for Disease Control and

Prevention, 20 June 2022, https://www.cdc.gov/diabetes/library/features/role-of-fiber.html#:~:text=Specifically%2C%20fiber%20can%20help%3A,sugar%20in%20your%20target%20range.

[12] Flores, Ashley C, et al. "Prospective Study of Plant-Based Dietary Patterns and Diabetes in Puerto Rican Adults." *The Journal of Nutrition*, U.S. National Library of Medicine, 3 Dec. 2021, https://www.ncbi.nlm.nih.gov/pmc/articles/PMC8643592/.

[13] Ghanbari-Gohari, Faezeh, et al. "Consumption of Whole Grains and Risk of Type 2 Diabetes: A Comprehensive Systematic Review and Dose-Response Meta-Analysis of Prospective Cohort Studies." *Food Science & Nutrition*, U.S. National Library of Medicine, 10 Mar. 2022, https://www.ncbi.nlm.nih.gov/pmc/articles/PMC9179146/#:~:text=Data%20regarding%20the%20association%20of,not%20reach%20a%20significant%20association.

[14] "IDF Diabetes Atlas 2021." *IDF Diabetes Atlas*, https://diabetesatlas.org/atlas/tenth-edition/. *The Impact of a Plant-Based Diet on the Prevention and Management of ...* https://cornerstone.lib.mnsu.edu/cgi/viewcontent.cgi?article=2108&context=etds.

[15] "Insulin Resistance." *Insulin Resistance | ADA*, https://diabetes.org/healthy-living/medication-treatments/insulin-resistance#:~:text=In%20response%20to%20the%20body's,elevated%20levels%20of%20circulating%20insulin.

[16] Jackson, Eric. "How to Eat a Plant-Based Diet." *UMMS Health*, 20 Oct. 2022, https://health.umms.org/2022/08/19/how-to-eat-a-plant-based-diet/#:~:text=Whether%20fresh%20or%20frozen%2C%20men,or%20spinach%20and%20Brussels%20sprouts.

[17] Jardin, Meghan A, et al. "Perspective: Plant-Based Eating Pattern for Type 2 Diabetes Prevention and Treatment: Efficacy, Mechanisms, and Practical Considerations." *Academic.oup.com*, https://academic.oup.com/advances/article/12/6/2045/6296092.

[18] Qian, Frank, et al. "Association between Plant-Based Dietary Patterns and Risk of Type 2 Diabetes." *JAMA Internal Medicine*, vol. 179, no. 10, 2019, p. 1335., https://doi.org/10.1001/jamainternmed.2019.2195.

[19] Satija, Ambika, et al. "Plant-Based Dietary Patterns and Incidence of Type 2 Diabetes in US Men and Women: Results from Three Prospective Cohort Studies." *PLoS Medicine*, U.S. National Library of Medicine, 14 June 2016, https://www.ncbi.nlm.nih.gov/pmc/articles/PMC4907448/.

[20] Tonstad, Serena, et al. "Type of Vegetarian Diet, Body Weight, and Prevalence of Type 2 Diabetes." *Diabetes Care*, U.S. National Library of Medicine, May 2009, https://www.ncbi.nlm.nih.gov/pmc/articles/PMC2671114/.

[21] "Type 2 Diabetes." *Mayo Clinic*, Mayo Foundation for Medical Education and Research, 19 Nov. 2022, https://www.mayoclinic.org/diseases-conditions/type-2-diabetes/diagnosis-treatment/drc-20351199.

Chapter 4: Veganism and Cancer

[1] National Cancer Institute. "What Is Cancer?" *National Cancer Institute*, National Institutes of Health, 11 Oct. 2021, www.cancer.gov/about-cancer/understanding/what-is-cancer

[2] "CPE Monthly: Skin Cancer and Nutrition." Today's Dietitian, www.todaysdietitian.com/newarchives/021115p50.shtml#:~:text=A%20plant%2Dbased%20diet%20rich%20in%20fruits%20and%20vegetables%20and,day%20for%20overall%20cancer%20prevention. Accessed 8 Dec. 2023.

[3] Sarnoff, Deborah. "Skin Cancer Information - the Skin Cancer Foundation." *The Skin Cancer Foundation*, 2018, www.skincancer.org/skin-cancer-information/

[4] "Latest Research on Skin Cancer | NutritionFacts.org." *Nutritionfacts.org*, nutritionfacts.org/topics/skin-cancer/. Accessed 9 Dec. 2023

[5] Yagi, K I, et al. "Malignant Melanoma of the Skin: A Case Report from the Eastern Region (Dammam Central Hospital), Saudi Arabia and Epidemiological Review." *Journal of Family & Community Medicine*, vol. 7, no. 2, 2000, pp. 67–71, www.ncbi.nlm.nih.gov/pmc/articles/PMC3437114/

[6] "CPE Monthly: Skin Cancer and Nutrition." *Www.todaysdietitian.com*, www.todaysdietitian.com/newarchives/021115p50.shtml#:~:text=A%20plant%2Dbased%20diet%20rich%20in%20fruits%20and%20vegetables%20and. Accessed 9 Dec. 2023.

[7] Goldman, Diana. "I Beat Melanoma, Then I Went Plant-Based to Help Stay Cancer-Free." Forks over Knives, 19 Apr. 2018, www.forksoverknives.com/success-stories/i-beat-melanoma-then-i-went-plant-based/. Accessed 9 Dec. 2023.

[8] "Breast Cancer." *Physicians Committee for Responsible Medicine*, www.pcrm.org/health-topics/breast-cancer.

[9] American Cancer Society. "What Is Breast Cancer? | American Cancer Society." *Www.cancer.org*, 19 Nov. 2021, www.cancer.org/cancer/types/breast-cancer/about/what-is-breast-cancer.html

[10] "Bee Venom Shows Promise, but Needs More Study." *Www.uclahealth.org*, www.uclahealth.org/news/bee-venom-shows-promise-but-needs-more-study#:~:text=The%20peptide%20melittin%2C%20which%20is. Accessed 9 Dec. 2023.

[11] Kwon, Na-Yoen, et al. "Anticancer Activity of Bee Venom Components against Breast Cancer." *Toxins*, vol. 14, no. 7, 5 July 2022, p. 460, www.ncbi.nlm.nih.gov/pmc/articles/PMC9318616/, https://doi.org/10.3390/toxins14070460

[12] "Honeybee Venom Kills Aggressive Breast Cancer Cells." *Www.medicalnewstoday.com*, 7 Sept. 2020, www.medicalnewstoday.com/articles/honeybee-venom-kills-aggressive-breast-cancer-cells#Blocking-messages

[13] *Pennmedicine.org*, 2022, www.pennmedicine.org/cancer/types-of-cancer/breast-cancer/types-of-breast-cancer/her2-positive-breast-cancer#:~:text=When%20breast%20tissue%20has%20extra

[14] “Triple-Negative Breast Cancer | Details, Diagnosis, and Signs.” *Www.cancer.org*, www.cancer.org/cancer/types/breast-cancer/about/types-of-breast-cancer/triple-negative.html#:~:text=Triple%2Dnegative%20breast%20cancer%20(TNBC

[15] “Honeybees May Offer Hope for Advancing Breast Cancer Treatment.” Honeybees May Offer Hope for Advancing Breast Cancer Treatment, www.facingourrisk.org/XRAY/honeybees-offer-hope-to-those-battling-breastcancer

[16] Mayo Clinic. “Prostate Cancer - Symptoms and Causes.” Mayo Clinic, Mayo Clinic, 14 Dec. 2022, www.mayoclinic.org/diseases-conditions/prostate-cancer/symptoms-causes/syc-20353087

[17] Zaragoza-Martí, A, et al. “Evaluation of Mediterranean Diet Adherence Scores: A Systematic Review.” BMJ Open, U.S. National Library of Medicine, 24 Feb. 2018, www.ncbi.nlm.nih.gov/pmc/articles/PMC5855302/.

[18] Tantamango-Bartley, Yessenia, et al. “Vegetarian Diets and the Incidence of Cancer in a Low-Risk Population.” Cancer Epidemiology, Biomarkers & Prevention : A Publication of the American Association for Cancer Research, Cosponsored by the American Society of Preventive Oncology, U.S. National Library of Medicine, Feb. 2013, www.ncbi.nlm.nih.gov/pmc/articles/PMC3565018/.

[19] Tantamango-Bartley, Yessenia, et al. “Are Strict Vegetarians Protected against Prostate Cancer?” *The American Journal of Clinical Nutrition*, vol. 103, no. 1, 11 Nov. 2015, pp. 153–160, https://doi.org/10.3945/ajcn.114.106450

Chapter 5: Alzheimer’s Disease

[1] “Alzheimer's Disease.” *FDA*, 4 January 2022, https://www.fda.gov/consumers/minority-health-and-health-equity-resources/alzheimers-disease. Accessed 8 December 2023.

[2] "Dementia." *World Health Organization (WHO)*, 15 March 2023, https://www.who.int/news-room/fact-sheets/detail/dementia. Accessed 8 December 2023.

[3] "Alzheimer's disease - Symptoms and causes." *Mayo Clinic*, 30 August 2023, https://www.mayoclinic.org/diseases-conditions/alzheimers-disease/symptoms-causes/syc-20350447. Accessed 8 December 2023.

[4] Murphy, Paul, and Harry LeVine. "Alzheimer's Disease and the β-Amyloid Peptide." *NCBI*, 29 January 2010, https://www.ncbi.nlm.nih.gov/pmc/articles/PMC2813509/. Accessed 8 December 2023.

[5] "What Happens to the Brain in Alzheimer's Disease?" *National Institute on Aging*, 16 May 2017, https://www.nia.nih.gov/health/what-happens-brain-alzheimers-disease. Accessed 8 December 2023.

[6] Medeiros, Rodrigo, et al. "The Role of Tau in Alzheimer's Disease and Related Disorders." *NCBI*, https://www.ncbi.nlm.nih.gov/pmc/articles/PMC4072215/. Accessed 8 December 2023.

[7] "Alzheimer's disease - Symptoms and causes." *Mayo Clinic*, 30 August 2023, https://www.mayoclinic.org/diseases-conditions/alzheimers-disease/symptoms-causes/syc-20350447. Accessed 8 December 2023.

[8] "Alzheimer's Facts and Figures Report." *Alzheimer's Association*, https://www.alz.org/alzheimers-dementia/facts-figures. Accessed 8 December 2023.

[9] Qiu, C., et al. "Epidemiology of Alzheimer's disease: occurrence, determinants, and strategies toward intervention." *NCBI*, 11 June 2009, https://www.ncbi.nlm.nih.gov/pmc/articles/PMC3181909/. Accessed 8 December 2023.ed September 22, 2023, from https://www.ncbi.nlm.nih.gov/pmc/articles/PMC3181909/

[10] Kim, Scott YH. "The ethics of informed consent in Alzheimer disease research." *NCBI*, 24 May 2011,

https://www.ncbi.nlm.nih.gov/pmc/articles/PMC3475518/. Accessed 8 December 2023.

[11] Alzheimer, Alois. "Impact of Alzheimer's Disease on Caregivers in the United States." *NCBI*, 20 August 2022, https://www.ncbi.nlm.nih.gov/pmc/articles/PMC9392839/. Accessed 8 December 2023.

[12] "What Do We Know About Diet and Prevention of Alzheimer's Disease?" *National Institute on Aging*, 20 November 2023, https://www.nia.nih.gov/health/what-do-we-know-about-diet-and-prevention-alzheimers-disease. Accessed 8 December 2023.

[13] "MIND Diet | The Nutrition Source | Harvard T.H. Chan School of Public Health." *Harvard T.H. Chan School of Public Health*, https://www.hsph.harvard.edu/nutritionsource/healthy-weight/diet-reviews/mind-diet/. Accessed 8 December 2023.

[14] Smith, PJ, and JA Blumenthal. "Dietary Factors and Cognitive Decline - PMC." *NCBI*, 1 March 2016, https://www.ncbi.nlm.nih.gov/pmc/articles/PMC4758517/. Accessed 8 December 2023.

[15] Broom, G., et al. "The ketogenic diet as a potential treatment and prevention strategy for Alzheimer's disease." *PubMed*, April 2019, https://pubmed.ncbi.nlm.nih.gov/30554068/. Accessed 8 December 2023.

[16] Ownby, R. L., et al. "Depression and Risk for Alzheimer Disease - PMC." *NCBI*, 26 December 26, https://www.ncbi.nlm.nih.gov/pmc/articles/PMC3530614/. Accessed 8 December 2023.

[17] Brommelhoff, Jessica A., et al. "Depression as a Risk Factor or Prodomal Feature for Dementia? Findings in a Population-Based Sample of Swedish Twins." *NCBI*, 20 July 2009, https://www.ncbi.nlm.nih.gov/pmc/articles/PMC2713179/. Accessed 8 December 2023.

[18] Byers, Amy L., and Kristine Yaffe. "Depression and Risk of Developing Dementia - PMC." *NCBI*, 16 April 2012,

https://www.ncbi.nlm.nih.gov/pmc/articles/PMC3327554/. Accessed 8 December 2023.

[19] Saiz-Vazquez, Olalla, et al. "Depression as a Risk Factor for Alzheimer's Disease: A Systematic Review of Longitudinal Meta-Analyses." *NCBI*, 21 April 2021, https://www.ncbi.nlm.nih.gov/pmc/articles/PMC8122638/. Accessed 8 December 2023.

[20] Juszcyk, Gabriela, et al. "Chronic Stress and Oxidative Stress as Common Factors of the Pathogenesis of Depression and Alzheimer's Disease: The Role of Antioxidants in Prevention and Treatment." *NCBI*, https://www.ncbi.nlm.nih.gov/pmc/articles/PMC8470444/. Accessed 8 December 2023.

[21] Gallagher, Damien. "Depression and risk of Alzheimer's dementia: A longitudinal analysis to determine predictors of increased risk among older adults with depression." *NCBI*, 9 May 2018, https://www.ncbi.nlm.nih.gov/pmc/articles/PMC6101977/. Accessed 8 December 2023.

[22] Teleanu, Raluca Ioana, et al. "Neurotransmitters—Key Factors in Neurological and Neurodegenerative Disorders of the Central Nervous System." *NCBI*, 23 June 2022, https://www.ncbi.nlm.nih.gov/pmc/articles/PMC9180936/. Accessed 8 December 2023.

[23] Botto, Rossana, et al. "Anxiety and depression in Alzheimer's disease: a systematic review of pathogenetic mechanisms and relation to cognitive decline." *NCBI*, 23 April 2022, https://www.ncbi.nlm.nih.gov/pmc/articles/PMC9213384/. Accessed 8 December 2023.

[24] Gorelick, Philip B., et al. "Vascular Contributions to Cognitive Impairment and Dementia: A Statement for Healthcare Professionals From the American Heart Association/American Stroke Association." *NCBI*, 21 July 2011, https://www.ncbi.nlm.nih.gov/pmc/articles/PMC3778669/. Accessed 8 December 2023.

[25] Tamashiro, Kellie LK. “The Brain on Stress: Insight from Studies Using the Visible Burrow System.” *NCBI*, 1 July 2016, https://www.ncbi.nlm.nih.gov/pmc/articles/PMC4777966/. Accessed 8 December 2023.

[26] Anacker, Christoph, et al. “The glucocorticoid receptor: Pivot of depression and of antidepressant treatment?” *NCBI*, April 2011, https://www.ncbi.nlm.nih.gov/pmc/articles/PMC3513407/. Accessed 8 December 2023.

[27] Sanchez, Raquel. “High Cortisol and the Risk of Dementia and Alzheimer's Disease: A Review of the Literature.” *NCBI*, 1 March 2019, https://www.ncbi.nlm.nih.gov/pmc/articles/PMC6405479/. Accessed 8 December 2023.

[28] Lagarto, L. “Behavioral and Psychological Symptoms of Dementia - PMC.” *NCBI*, 7 May 2012, https://www.ncbi.nlm.nih.gov/pmc/articles/PMC3345875/. Accessed 8 December 2023.

[29] Dubois, Bruno, et al. “Preclinical Alzheimer's disease: Definition, natural history, and diagnostic criteria.” *NCBI*, 14 March 2019, https://www.ncbi.nlm.nih.gov/pmc/articles/PMC6417794/. Accessed 8 December 2023.

[30] Beurel, Eleonore, et al. “The Bidirectional Relationship of Depression and Inflammation: Double Trouble.” *NCBI*, 22 July 2021, https://www.ncbi.nlm.nih.gov/pmc/articles/PMC7381373/. Accessed 8 December 2023.

[31] Clapp, Megan, and Emily Wilen. “Gut microbiota's effect on mental health: The gut-brain axis.” *NCBI*, 15 September 2017, https://www.ncbi.nlm.nih.gov/pmc/articles/PMC5641835/. Accessed 8 December 2023.

[32] Minakawa, Eiko N. “Bidirectional Relationship Between Sleep Disturbances and Parkinson's Disease.” *NCBI*, 18 July 2022, https://www.ncbi.nlm.nih.gov/pmc/articles/PMC9342689/. Accessed 8 December 2023.

[33] Whitten, Lori A. "Functional Magnetic Resonance Imaging (fMRI): An Invaluable Tool in Translational Neuroscience." *NCBI*, December 2012, https://www.ncbi.nlm.nih.gov/books/NBK538909/. Accessed 8 December 2023.

[34] Remes, Olivia, et al. "Biological, Psychological, and Social Determinants of Depression: A Review of Recent Literature." *NCBI*, 10 December 2021, https://www.ncbi.nlm.nih.gov/pmc/articles/PMC8699555/. Accessed 8 December 2023.

[35] Ellouze, Ines, et al. "Dietary Patterns and Alzheimer's Disease: An Updated Review Linking Nutrition to Neuroscience." *NCBI*, 19 July 2023, https://www.ncbi.nlm.nih.gov/pmc/articles/PMC10384681/. Accessed 8 December 2023.

[36] Wang, Richard C., and Zhixiang Wang. "Precision Medicine: Disease Subtyping and Tailored Treatment." *NCBI*, 28 July 2023, https://www.ncbi.nlm.nih.gov/pmc/articles/PMC10417651/. Accessed 8 December 2023.

[37] Wells, J.L., and S.H. Pasternak. "Case Report of a 63-Year-Old Patient With Alzheimer Disease and a Novel Presenilin 2 Mutation." *NCBI*, April 2019, https://www.ncbi.nlm.nih.gov/pmc/articles/PMC6554015/. Accessed 8 December 2023.

[38] Zhu, L., et al. "Case of early-onset Alzheimer's disease with atypical manifestation." *NCBI*, 28 January 2021, https://www.ncbi.nlm.nih.gov/pmc/articles/PMC7845665/. Accessed 8 December 2023.

[39]Um, Yoo Hyun, et al. "A Case Report of a 37-Year-Old Alzheimer's Disease Patient with Prominent Striatum Amyloid Retention." *NCBI*, July 2017, https://www.ncbi.nlm.nih.gov/pmc/articles/PMC5561413/. Accessed 8 December 2023.

[40] Chin, SS. "Alzheimer's disease." *PubMed*, 3 October 2001, https://pubmed.ncbi.nlm.nih.gov/14602942/. Accessed 8 December 2023.

[41] Jia, Jianping, et al. “A 19-Year-Old Adolescent with Probable Alzheimer's Disease.” *PubMed*, 2023, https://pubmed.ncbi.nlm.nih.gov/36565128/. Accessed 8 December 2023.

Made in the USA
Monee, IL
08 November 2026

68002500R00080